The God Clock

a new look at the passage of time

By Thomas Morrell

The God Clock
a new look at the passage of time

Edited by Karen Holleran

Library of Congress Cataloging-in-Publication Data is available.
The God Clock, a new look at the passage of time by Thomas Morrell

ISBN 978-0-96651-632-6

Dedication

To my friends:

To Karen for her great editing and support

To Casey for reading my drafts and providing the sub title

To all previous authors who have thought outside the box to bring us to this understanding at this point in time.

And to the Native Peoples around the Earth, who have kept the truth alive.....

Table of Contents

Introduction

Just as some of our most brilliant minds have pondered great mathematical questions, there is an even greater one that has not been completely answered yet – what is Time? Some believe that it may not exist at all and that it is simply an illusion, while others believe it is real, but can change and think it is possible that it can bend. As our technology expands and we looker deeper out into the Universe, our concepts of time are evolving and changing. We race forward to understand time using grand and complex theories, but some of the real missing pieces of the puzzle, may actually be both in the past and surprisingly, within us. Again, we are evolving and our thoughts and understanding growing, while the Native American beliefs have remained the same – "everything is a circle" and this book may show they have been right, all along.

Within the pages of this book, the author takes us on a journey back into ancient calendars, structures and systems, to not only understand time itself, but just as important – our journey through it, as well. Looking into the past, we will see that time is not only divided into specific increments, such as days, months, seasons and years, but also in specific periods, cycles, proportion and all within a circle. Once we understand the make up of time, later the author proposes the Mayan Time Grids, even larger blocks of time and the influence of those increments upon us.

Did you know, in 1982 we started the "Information Age" that will last for 52 years?

We will try to understand, or comprehend what Time is made of, its basis; its essence, or matrix and look at ourselves as possible time pieces, operating perfectly within it. To see Time literally as a fabric and how everything, including us, is woven into it, showing that everything is a circle and that we are all connected. If Time really is a circle, will what has happened before, happen again?

An even deeper question, where does Time come from? The author suggests the possibility of a grand mechanism in which all time unfolds known as "The God Clock," which could very well be the consciousness, or soul of the universe.

After reading this book, it is possible you will never look at the Universe, our Solar System, the Earth, our life on it, yourself and how you connect to others and Time in the same manner ever again.

Chapter One

The Time Code
The Precise Unfoldment of Time

Ecclesiastes

To every thing there is a season, and a time to every purpose under the heaven:

A time to be born, and a time to die; a time to plant, and a time to pluck up that which is planted;

A time to kill, and a time to heal; a time to break down, and a time to build up;

A time to weep, and a time to laugh; a time to mourn, and a time to dance;

A time to cast away stones, and a time to gather stones together; a time to embrace, and a time to refrain from embracing;

A time to get, and a time to lose; a time to keep, and a time to cast away;

A time to rend, and a time to sew; a time to keep silence, and a time to speak;

A time to love, and a time to hate; a time of war, and a time of peace.

As we go through life, there is one constant issue that we deal with every day – TIME. Time to get up, time to go to work, time to eat, time for our favorite TV show, time for a snack, time to go to bed., time for a change, or maybe when we meet someone special in that wonderful moment in time.

We are born, we live and at some point we die – in Time.

We can perceive time both slow and fast. When we are involved in something boring, time seems to unfold very slow. The older we get time seems to go by faster and faster and sometimes when we are involved in something profound, time seems to even stand still. Although real, these are just perceptions of time as it has been precisely unfolding since – the beginning of time.

A Journey Through Time

Back in the 1990's, research was done into an old symbolic astronomical system based on the mathematics and symbology behind an ordinary deck of playing cards in relation to the squaring, or quadrature of a circle's diameter. Through this research a Star/Time Code was re-discovered, a mathematical sequence of numbers that relates closely to the orbits of Saturn, the Earth, the Moon and seasons of those motions, that will be mentioned in this book. The work published on this and the old symbolic astronomical system that works within it is known as "The Ancient Book of Time." We will go into more detail of the system in a later chapter as it relates to many mysteries and symbolism around the world. The increments of time known back then regarding the symbolic astronomical system was a 91 year cycle, a 7 year cycle and of course a Yearly cycle. Each cycle was divided into 7 planetary periods and each period assigned a planet in order from

Mercury through Neptune. The basis of the 7 Divisions or, Planetary Periods within time is founded on the relationship of a circle's diameter to its circumference and also hidden within the Masonic Symbol, which will be addressed in Chapter 5.

It was obvious that the symbolic astronomical system in its state at the time was not complete and was simply incorrect in many areas. Because of this, some techniques and interpretations were off, or being twisted into meaning something they did not, and is still being used in that manner in some circles today. Understanding the Star/Time Code may also point to inaccuracies within the Mayan and Aztec Calendars as we have a similar situation going on there – much of the information is missing and without it, some techniques and interpretations are flawed. Take the prediction of December 21st, 2012 as an example, the end of the Mayan Calendar, or the end of the world and/or major shifts and changes with/on Earth – didn't seem to happen. This will be explained in further detail and straightened out in the next chapter regarding the Mayan and Aztec Calendars.

The Cycles

A Yearly Cycle of course is based on the Earth's orbit around the Sun, but what are the 7 year and 91 year cycles based on? It takes approximately 29 years for the planet Saturn to make one orbit around the Sun, therefore it takes that long for Saturn to return to the point in the Zodiac it was about 29 years before. For example, Saturn was at approximately 5 degrees Sagittarius December 29th, 1985 and then on March 6th, 2015 it returned to that position.

To Everything There is a Season

In the Bible it states: “To everything there is a Season, a time and purpose under heaven”

Understanding the orbit of Saturn, its journey around the Sun and through the 12 Signs of the Zodiac, the Star/Time Code started to reveal itself. The Yearly cycle is of course based on the orbit of the Earth around the Sun, but again, what is the 7 year cycle based on? We have all heard of the 7 year itch and it is believed that the cells in the body are replaced every 7 years. Approximately every 7 years, the planet Saturn, transiting around the Zodiac, makes or forms a hard aspect, or angle to it’s location in one’s Astrological Natal Chart (a chart to display in a graphic format a picture of the heavens above at the moment a person was born) it creates a square (90 degrees), opposition (180 degrees), another square (90 degrees) and finally a conjunction (0 degrees, known as the Saturn Return.) The 7 year cycle then, is simply a quarter, or Season of Saturn’s motion.

Father Time
Linear vs Cyclic

Although not known before in modern times, this understanding of the 7 year cycle, uncovers a larger 28 year Saturn cycle within the Star/Time Code. This new 28 year Saturn Cycle will be divided into 7 Planetary Periods, Mercury through Neptune, just like in the other cycles of time. Each planet possesses certain attributes, influence, as well as rulership over specific areas and things. The remarkable and maybe most important component in this discovery of this cycle is that when you look into any astrological rulership reference material, or ancient mythology, the planet Saturn itself – governs TIME! – As we will see, Saturn is the key to understanding, and basis of the Star/Time Code.

In Roman Mythology Saturn is depicted as an older man with a long beard, signifying a person that has lived a long life, hence the saying "Father Time." Usually with such experience comes wisdom, which Saturn exemplifies and teaches. He was also portrayed in Greco Roman mosaics as a man turning the Zodiac wheel When the orbit or cycle of Saturn is completed, whether in one's Natal Astrological Chart, or in the Star/Time Code, it is considered the actual gaining of wisdom and experience. Saturn also represents structure and form -- the building of Time.

There is difference between the actual time it takes for Saturn to orbit the Sun, approximately 29 years and the Star/Time Code increment of Saturn of 28 years. We have to ask then, what is the difference between Linear/Astronomical Time (29+ year actual orbit of Saturn) that unfolds within the natural seasons and cycles in a seemingly straight line and Cyclic Time (28 year Star/Time Code duration of Saturn), time that unfolds in precise increments? It could be looked at in this manner Linear/Astronomical Time is more of the environment we are going through, while Cyclic Time is the path we are on – this will be explained more in the a later chapter as it may be a key in unlocking the mystery and true meaning of the Great Pyramid in Egypt.

Smaller Cycles

We now know what and where the 7 year Cycle comes from and have discovered the new, important 28 year Saturn Cycle, but where and what are the other pieces of the puzzle that are missing? In the Ancient Book of Time (the ancient astronomical system based on the mathematics and symbology behind the deck of playing cards in relation to the squaring, or quadrature of a circle's diameter) one calendar year is represented by the Number 364. There are 365.25 days in a year so the calendar is adjusted through December 31st and February 29th for everything to fall into place. The two days are divided in half and each half given to the day preceding and following that day – 365.25 is Linear time, 364 is Cyclic time. In many ancient cultures December 31st

was a day where no work was done, used as a day of celebration and was not counted as an actual day in the calendar. February 29th is just an extra day added each 4 years to get the calendar back in sync. Each year is divided into 4 Seasons – Spring, Summer, Fall and Winter, so when we divide a year (364 days) by 4 for the 4 seasons we get 91, the number of days in each season.

The Bigger Picture

We experience 4, 91 day Seasonal Cycles each year, which gives us a clue as to what the 91 year Cycle really is – a seasonal increment to a larger cycle of time, a 364 year cycle. What is extremely enlightening regarding this larger cycle of time is that when it is divided into each of the 7 planetary periods within it, each period is 52 years, an important increment of time in the Mayan Calendar. We will go more into the workings of this in later chapters, but to give you a glimpse into the relevance of this, redoing the calculations with new information, in 1982 (which is possibly the real end, or important shift in the Mayan Calendar) we started a 52 year Mercury Period – The Information Age.

In Mythology, Mercury is known as the winged messenger of the gods, ancient Greeks referred to him as Hermes and the Egyptians as Thoth. Mercury in astrological terms represents knowledge, communication, information, thought and quickness. The Ancient Book of

Time was originally referred to as The Book of Hermes and/or The Book of Thoth.

While there are other sub-cycles/planetary periods going on during this 52 year period, with the advancement of the internet, cell phones, 24 hour news, online classes, ebooks, etc., what better and more accurate way to describe what is going on, on a large scale around us? Literally, an explosion of information and communication, all ruled by Mercury.

With the new, larger 364 year cycle discovered, this gives us an understanding of two important increments of time – the basis and origin of the 91 year cycle in the Ancient Book of Time and the 52 year Mayan Cycle (which is really an increment in the division of a cycle, not a cycle unto itself) that is talked about, but never accurately explained, until now.

Back to the Smaller Cycles

So far we have a 364 year Cycle, a 28 year Saturn Cycle and a 364 day Solar Cycle, all with their respective Seasonal Cycles (91 years, 7 years, 91 days) within them, but there are smaller increments of time that we also experience here on Earth. The next important motion of a heavenly body smaller than the 91 day Seasonal Cycle would be the Moon. The Moon is Earth's only satellite and effects the tides of the oceans and the timing of New and Full

Moons is tracked and used by many today for planting, harvesting, or starting new projects and finish others. This is a 28 day Lunar Cycle and within this, 4 - 7 day Seasonal Weekly Cycles. Again – "to everything there is a season." Similar to Saturn's slight difference between its actual time of orbit vs the cyclic time in the Star/Time Code, the Moon does the same thing. The actual time it takes the moon to transit each sign of the Zodiac is approximately 29 days, the increment in the Star/Time Code is 28 days.

All of these Cycles represent the unfolding of time, but unlike others these are closely tied to the natural cycles and Seasons we experience here on Earth, as well as to 3 specific planetary bodies, Saturn, the Earth and the Moon. Now let's look even deeper into the magnificence and precision in this Star/Time Code.

A Linear Repeating Code

Even though there are even larger and smaller Cycles than the ones that were known and the new ones we have discovered, the 8 that we know of now are remarkable in their own right. First we have, in order, the 364 year Cycle, the 91 year Cycle, the 28 Saturn year Cycle and the 7 year Cycle. Then we have the same Code unfolding in days – 364 day Solar Cycle, the 91 day Seasonal Cycle, the 28 Day Lunar Cycle and the 7 day Weekly Cycle, a linear code that simply and beautifully repeats itself – 364....91....28....7....364....91....28....7.

Cycles within Cycles

There are many things going on within the Star/Time Code besides just the linear repeating code and they not only show how profound and powerful the Code is, but also gives it substance and validity.

364 is a very significant increment of measure as the integers add up to the Number 13. We will get into the importance of the Number 13 shortly, but for now just know it is important, especially in regards to completion. Let's look at the remarkable Cycles hidden within. In a 364 years, there will be 364, 364 day Cycles, in 91 years there will be exactly 364, 91 day Seasonal Cycles, in 28 years there will be exactly 364, 28 day Lunar Cycles and in 7 years there will be exactly 364, 7 day Weekly cycles.

364Years – 364, 364 day Cycles
91Years – 364, 91 day Cycles
28 Years – 364, 28 day Cycles
7 Years – 364, 7 day Cycles
1 Year – 364 days
(the Year being adjusted with 12/31 and 2/29)

Grand Precision and Circles

The Star/Time Code is showing us that Time is not random and that it unfolds in a very precise sequential order on a linear repeating path, but, also in precise cycles that are linked to larger grand cycles that we are all part of and connected to. This could also point to and validate what the Native Americans believe – that everything is a Circle.

The Number 13

Most think of the Number 13 as an unlucky influence, but in reality, it is an important number and represents something special in relation to the unfolding of Time – it stands for completion. Tarot Cards are usually associated with Fortune Telling games and tricks, while in reality they have a deeper connection to the structure of our Solar System, the ancient Hebrew/Chaldean alphabet and even the Masonic Symbol, which we will get into in a later chapter. This is mentioned as the 13th Card of the Major Arcana of the Tarot is the Death Card and symbolizes Death and Rebirth, the ending of something and the beginning of another.

While there are endings in our everyday lives (that happen precisely when they are suppose to), in order for a Cycle to be completed, it must go through the 4 seasons which are: Creation, Growth, Harvest and Death. A Solar Year Cycle in the Star/Time Code goes through 364 days and the integers 3-6-4 add up to the Number 13 and let's not forget there are 364 smaller day cycles in their respective, larger 364, 91, 28 and 7 Year cycles, all making completions in their own small way, in time.

In a Solar (Yearly) Cycle there will be exactly 13, 28 day Lunar Cycles. In a 91 day Season Cycle there are 13 weeks and also 7 Planetary Periods (divisions of time) of 13 days. In the regular deck of 52 playing cards there are 4 Suits with 13 cards in each suit and when all the spot values of the cards are added together in all 4 suits (1 - 13) the total is 364.

The simple reason that the Number 13 is the number of completion is that when you add the integers together (1+3) the sum is 4, the number of Seasons in a Cycle. In order to complete any Cycle you must go through the 4 Seasons within it and not necessarily in a means as defined by temperature – Spring/warm, Summer/hot, Fall/cool and Winter/cold, but again, by action/evolvement – Creation, Growth, Harvest and Death. An interesting note, when you add all the numbers up in one block of time (364 + 91 + 28 + 7) whether it be years or days, the total is 490. And when you add these integers together (4 + 9 + 0) again we get 13.

Over and Over

This reveals that Time is not only unfolding in the precise sequential, linear and in a cyclic manner, but that there are also large and small increments of completion that happen in a specific seasonal – creation (Spring), growth (Summer), harvest (Fall) and death (Winter) environment, along with the 7 divisions and planetary periods within each of the cycles – and – over large and small periods of time these cycles and influences repeat themselves like a beautiful symphony. Again, everything is a circle.

Using the Star/Time Code we can then map out time in specific cycles along with its Seasonal and Planetary influences to better understand the Past, evolve in the Present and glimpse into the Future. As mentioned before, there are even larger blocks of time but they may be misunderstood or even mislabeled, creating the possibility that parts of the Mayan and Aztec Calendars could be off, misinterpreted or explain what has been missing. We will get into that in the next chapter.

The Telios Proportions

"In the science of the arts, beginning in 1898, Dr Landone studied 153 systems of proportions and after 38 years of research, he announced in 1936 his *Re-discovery of the Telios – the Basis of the Canon of Polykleitos"* Robert Collier on Dr. Brown Landone

We have laid out increments of time based on motions of specific planets/heavenly bodies and how the time, or the numbers associated with these form a unique and remarkable code with linear, sequential and cyclic properties, but there is even more to the story – proportion – more specifically the Telios Proportions. Telios is the Greek word for perfect, so Telios Proportions signifies Perfect Proportions.

Dr. Brown Landone states, "All things of nature appear to be created on the basis of the Telios Proportions." – including Time itself. The base numbers of the Telios are 1,

4, and 7 and as pointed out, we experience 1 cycle with 4 seasons, divided into 7 periods of time. In music there are 4 basic tones, 7 divisions on the major scale and 13 intervals in the minor scale, but it can also be interpreted in another aspect – Music or Harmony of the Spheres is an ancient philosophical concept regarding the proportions in the movements of celestial bodies with the Sun, Moon and planets seen a form of music. This isn't music we actually hear, or listen to, but more of a harmonic, mathematical, or even a spiritual perception.

According to Dr. Brown Landone, there are 7 sacred temples around the world that are built specifically to Telios Proportions, one of which is the King's Chamber in the Great Pyramid. He goes on to say "In the structure of the Pyramid in Gizeh the Telios is so dominant that we are forced to believe that it was intentionally used to symbolize and record knowledge of the past and prophecies of the future." He mentions the locations of the 7 Key Temples: Tiahuanaco (Bolivia), Cathay (China), Mayaland (Chichen Itza, Mexico), Euleusis (Greece), Thibet (Himalayas), The Great Pyramid of Egypt and one being built in the East Andes Mountain Range around 1940?

Within these 7 Temples you can draw circles of Telios proportions and where these lines intersect, it reveals specific dates of events in regards to the evolvement of man.

According to Dr. Brown Landone this information and how it is ascertained is sacred and only known to those initiated into some high order.

Dr Brown Landone was born on March 6, 1847 on a boat at sea and died at the age of 98 on October 10, 1945 in Winter Park, Florida. Listed are some of his credentials: Educated in America, England and France. Studied at and received a Doctorate of Divinity from Episcopal High Church in England. Later studied and received a medical degree and had his own medical practice. Was the Editor in Chief of a History of Civilization – Victories of Scholarship, Science and Art (1909.) In Paris, before 1914, Dr Landone was the Director General of the Institute of the Science of Arts in France.

Time Itself and Our Time

We know Time is unfolding in precise cycles, planetary divisions and now in perfect proportion, as well. In a later chapter we will take a look at specific increments of Time that happened in the not to distant past and current influences to understand this better. While the influence of Time itself is happening over large blocks or increments of time down to each single moment, we to, are following our own path in this life – as defined by our Time. We also experience our own personal cycles of time, planetary divisions and in perfect proportion.

The Body is a Temple

From Corinthians 6:19-20, "Do you not know your body is a **Temple** of the Holy Spirit"

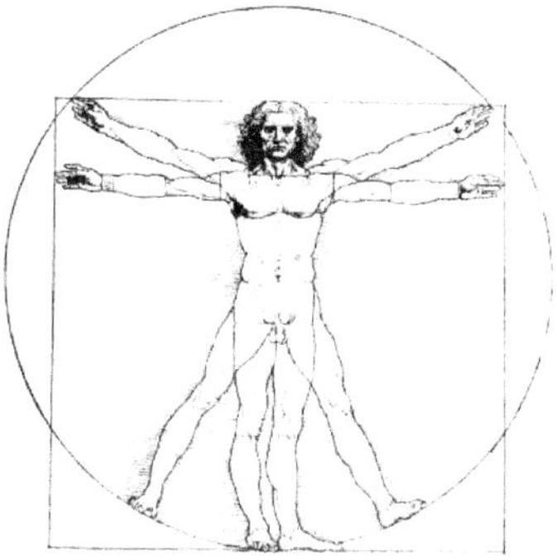

Just as within the 7 sacred Temples built around the World, circles can be drawn within to Telios Proportions to reveal, or map out the timing of significant events in the evolution of man, since the human body is also a Temple and designed to Telios (perfect) proportions, cannot circles be drawn within, or from us to also reveal, or map out our path in this life? As we will see in the Chapter 4, the ordinary deck of playing cards mathematically reveals that map, in symbols, planetary divisions and in perfect proportion as defined by the Telios.

Something Truly Remarkable

All of these blocks, or circles of time are constantly unfolding, perfectly, then during one moment, on a specific day, during a particular week that is part of a lunar cycle, during a certain season, in a given numbered year – you are born. That moment/day is your marker and your circle of life begins. Now your circles, or cycles are unfolding; the 364 year, 91 year, 28 year, 7 year, One Year, 91 day, 28 day and 7 day cycles, all separate, but together at the same time and when specific influences line up, or circles intersect, significant events will transpire in your life. While your own circles of life are unfolding, along and within the circle of Time itself, other's circles of life are unfolding – AND – their circles will intersect with others! Where our circles intersect with other's circles represents people coming in and out of our lives, some for a lifetime and some for just a brief moment. Some for work, some for love, some to purchase a car, or get a cup of coffee. From the tragic shooting in Ferguson, Missouri that happened on August 9th, 2014, using The Ancient Book of Time System that will be explained in Chapter 4, it clearly illustrates not only what was going on in Michael Brown's and Darrell Wilson's lives at the time of the event, but the intersection between them!

The Mechanism of Time
The God Clock

We all have our vehicles on which we use to travel within our lifetimes, whether it is our feet, a bike, a car, or maybe even an aircraft. So the question has to be asked, what is the vehicle used by Time on its journey, or path?

The essence, or vehicle on, or in which time unfolds – might be the fabric of the universe that holds all things together and/or be like a magnificent, precise machine in which all circles of time: time itself, persons, things, places, events and even ideas are all unfolding, each in their own, separate, seemingly linear, sequential, cyclic, symmetrical and proportional way, yet all are working/meshing perfectly together, some intersecting, some overlapping, some just near and some not even close, but still connected. It is multi-dimensional, emanating in all directions and on many levels – physical, emotional, spiritual and ethereal. Circles from the Past, intersecting with circles in the Present and in turn connecting with circles in the Future. Lets add to that the Universe, all its galaxies, stars, other solar systems and possible life, again circles of time all unfolding, as they should. Just think of the complexity and shear numbers of circles unfolding in the Universe at this moment from the largest increment of time down to each second, in what ever essence, fabric or machine it is, in a perfect, precise manner, called – The God Clock.

Again, just think about what happened between Michael Brown and Darren Wilson and the tragic event that transpired. You have two people whose circles of Time are unfolding separately, then in one moment their circles intersect and not only theirs, but others around them and not only just them but the city of Ferguson, Missouri as well, which also has its own circles of time unfolding and all of those circles intersecting in a specific moment in time itself – incredible – but that is just a small glimpse of how The God Clock works.

Larger Intersections

Besides people having their circles of time unfolding and intersecting for good and bad, when countries and governments are formed, their circles start to unfold as well and theirs will be intersecting with the population and leaders of that country. As we go through time, different countries circles will connect, or intersect with others for good, or bad. Some will make alliances, others will become enemies, some will be for a long time, others just for short periods and the depending on each country's path at the time of the intersection, and the governing planet within that time (and other environmental factors), it would show the nature, or potential of that connection. What if there is life out there in the universe besides life on Earth and their circles of time one day intersect with the circles of time of the Earth – for good or bad?

Now vs Then

In today's modern society, even with our atomic clocks and quartz watches that are supposedly so accurate we really don't track time in any meaningful, deeper way. Oh we set the alarm to get up for work, we watch the clock to see how long it is before we get off work, time for lunch, time for dinner, time to go out, what time does the movie start, how long does it take us to go from point A to Point B, how long is the flight, how long do we have to wait in line, or how long do we have to wait on the phone for the next available representative?

Unlike modern man, the ancients and indigenous peoples of the past knew that there was something special about time and they went to great lengths to record and track it. They knew that there were cycles that unfolded, along with seasons and smaller increments within them and that they repeated themselves and by knowing this, they could plan for significant events and/or influences to come. Whether it was weather patterns such as a drought, fortunate times to plant and times to harvest.

However, modern man knows very little about time on a deeper level and is simply reacting to it in a very uneducated way. Looking at the world today under its current influences, it would appear as a whole we are making poor decisions and choices within the circles of time and the outcomes, many times, do not seem to be fortunate, but, again "to everything there is a season, a time and a purpose under heaven."

The difference between us and them is, they revered time and understood it much more than we do today and therefore used it to understand both themselves and the world around them, how they were connected to it and could try and make good choices and actions within it to achieve the most favorable outcome with the circumstances and influences past, present, or to come – togo with the flow, so to speak. By tracking time over hundreds, maybe even thousands of years, they learned from it – some in simple ways and other such as the Mayans and Aztecs in a very precise detailed manner.

We think we are living our own, separate lives on a straight, as time goes by path. The ancients and indigenous peoples not only thought, but knew – everything is a circle and that we are all connected.

When Things in Time End

Many things in our lifetimes come to an end – relationships end, jobs end, sporting events end (sometimes in a tie), wars end, days, months, and years end – even our lives, at some point come to an end. Many predicted the end of the Mayan Calender on December 21st, 2012 and that it would either be an actual end to the world, or an ending, and/or significant shift in the world and mankind, possibly on different levels, both spiritual and physical?

As far as the end of the Mayan calendar, (we will deal with it in more detail in the next chapter) most likely they knew of The God Clock (but probably didn't call it that) and simply quit putting the calendar out. How far into the future do you have keep it going, especially if your civilization isn't even around any more? If there wasn't going to be a human sacrifice at 12:00 PM on December 22nd, 2012, why put the calendar out? One theory was that the great 52 year cycles synced up with a larger 26,000 year increment of time on that day, but by calculations made using the Star/Time Code, we will show that time may unfold differently and the big shift, or changed happened within another point in time. Once the frame work is in place on how to mark time and interpret it, they knew down the line an intersection in time would be made and someone would figure it out and put out the new (insert the year here) calendar complete with cute puppies in it.

We experience endings in our personal lives, but knowing that we are on a path within the circles of time, one has to wonder if we hold onto an ending to long, are we then trapped inside that circle? Are we continually learning the lesson until we finally let go and then experience an intersection with another circle, finally taking us in a new direction, but still on our path within The God Clock?

Now that we have a better understanding on how time might unfold and the possibility of the existence of The God Clock, let's continue our interesting journey through time.

The End of the Time Chapter......

Chapter Two

The Mayan and Aztec Calendars

In Chapter 1 we learned how and why time unfolds in the way and manner it does. Because of the importance placed upon the Mayan and Aztec Calendars and events, such as December 21st, 2012, in this chapter, we are going to delve into this area. As most know, much of this important information was destroyed by the invasion of the Spaniards and what we have today is a partial working theory of what it actually was and how it was used. After showing one of the accepted versions of the Mayan Calendar, we will show a theory that was published in the 1930's on the Aztec Calendar that really demonstrates its true workings.

Later in Chapter 6 we are going to go into much greater detail into blocks of time and their influence which will be known as the Mayan Time Grids. For now, let's take a look at why we ended up at this point in time.

The Great Mayan Calendar

Who hasn't heard of the Mayan Calendar? According to all that has been written on it and dozens of ancient mystery shows on TV about it, it has to be one of the most incredible time pieces ever invented – or is it? The problem with ancient knowledge and information of any kind might not be whether it is accurate or not, but are we understanding/interpreting it correctly? The bigger problem with the Mayan Calendar is that not only are we probably interpreting it incorrectly, but as mentioned, due to the invasion into their culture of the Spaniards, much of the information and knowledge was destroyed. The version we are getting today is only a partial one and when that is the case, much of it is creatively filled in that may sound good, but when tested, usually doesn't pan out. The same problem happened with the Symbolic Astronomical System based on the deck of playing cards, much of the System was missing, so techniques were created from flawed interpretations to try and fill in the blanks. With no one to question any of this, it was considered accurate and almost mystical. And just as with the Mayan Calendar, many of these flawed techniques are still being used today. Add to that a proposed sense of New Age enlightenment only known to those to whom this ancient wisdom has been revealed, then you can really get some mis-information.

The Calendars

In the book, The Book of Destiny by Carlos Barrios, a Shaman and member of the Mayan Elder Council, he reveals calendars based on both days and years. The basic day calendars are a 260 day calendar, a 360 day calendar and a 400 day calendar. The 260 day calendar is said to be based on the orbit of Venus and the gestation period of a woman's pregnancy. The 360 day calendar is an astronomical one and along with it 5 days are added to fill out the 365 day year. (Nothing is to be done on these 5 days as they are considered malefic, or bad.) The 400 day calendar is given no basis other than the fact a full cycle is completed every 5200 days and may even be a 20 month calendar of 20 days (20 x 20 = 400)?

What seems to be very prominent within the Mayan and Aztec Calendars and our current interpretation of them, is a 20 increment system, by 20 glyphs, or symbols displayed on their calendric illustrations. These are interpreted as days and years and are the basis of the 260 day and 360 day calendars. In the case of the 260 day calendar, it is divided into 13 months of 20 days (13 x 20 = 260) and the 360 day calendar is divided into 18 months of 20 days (18 x 20 = 360.)

Again, the 260 day calendar is said to be based on the orbit of the Planet Venus and the gestation period of a woman's pregnancy. It actually takes Venus approximately 224.70 days to orbit the sun and a woman's gestation period is more like 270 days, so it is not even close. Remember, the Planet Saturn is given the governance of Time, not Venus. The astronomical calendar of 360 days with its added 5 days works, but Barrios also talks of a 260 year calendar that is based on the 360 day calendar and that it completes its cycle every 93, 600 days (260 x 360 = 93,600), so where are the 5 days added every year in regards to that? No mention of any adjustment is made, where as The Ancient Book of Time is consistent from the smallest increment to the largest. Barrios also mentions a 52 year calendar based on the cluster of 7 stars known as the

Pleiades, but again any real detail, or basis is missing. The Pleiades may actually be an important point in the galaxy that our circles have connected with in the ancient past and could again in the future.

The Mayan Version of 2012

According to Carlos Barrios, on December 21st, 2012 the Earth aligned with the center of the galaxy, which happens every 26,000 years. Barrios also mentions a 26,000 year calendar divided into 5, 5200 year cycles and on December 21st, 2012 we entered the 4th of 5 of these cycles. You would think with such an important astronomical event that happens every 26,000 years that it would be the time marker, the ending and beginning of a new long period instead of the ending of a 4th 5200 year cycle and the start of the final, or 5th 5200 year cycle. Another interesting point he makes is that the 4th 5200 year cycle that ended on December 21st, 2012 was governed by the element of Water, Water being the reigning sign for that time and according to Barrios, governs knowledge. In astronomical circles, the element of Water is associated with feelings, nurturing and emotions. Air is the element associated with knowledge. The question has to be asked, why does the Element of Water represent something in other cultures, or systems, but in the Mayan System it represents something completely different?

Along with the 2012 event Barrios mentions another cycle that has 2 smaller cycles within it – one happening from 1987 to 1992 and another from 1992 to 2001. The first 5 year period was, according to Barrios, an intense time and the second 9 year gestation period of some kind. Where do these come from and what is their basis?

The Problem

Besides the fact we know much information is missing from its destruction by the Spaniards in regards to the Mayan Calendars, as just illustrated it is really just a mixed up compilation of random increments and cycles with no real continuity. And as mentioned in the beginning of this chapter, with so much missing, they are most likely interpreting it incorrectly and just trying to come up with something that sticks. December 21st, 2012, for example, was supposed to be a great time of change, or shift with mankind – was it? Some even say the Mayan Calendar actually ended on that day, while Carlos Barrios says it is the start of another important 5200 year cycle, so who is right? In chapter 6 we are going to get into specific time makers and their meanings and influence in what we will call the Mayan Time Grids, but for now, let's look at another view on the Mayan and Aztec Calendars.

The Solution
Another View on the Calendars

In the National Astrological Journal published in the 1930's are two very interesting articles on the Aztec Calendar by Elbert Benjamine. Elbert Benjamine was an Astrologer (born in Adel, Iowa in 1882) who was the Head of the Brotherhood of the Light, based in California. Elbert Benjamine studied at Iowa State University to become a Naturalist, was the President of the Southern California Nature Club for 25 years and was responsible for the construction of a bird sanctuary in Los Angeles.

The Aztec and Ancient Calendars

In the first article, according to Elbert Benjamine, down through time the Ancients have tracked and documented the movements of heavenly bodies in order see and determine their influence upon earth. The first example is the Anu Enlil series, a major series of 68 or 70 stone tablets, written almost 3000 years ago, on which were written over 6500 to 7000 omens, of which it is believed to span a thousand years almost unbroken, that interpreted a wide variety of celestial phenomenon relative to the king and state. The later recognition of the Chaldeans was founded upon the accuracy secured by these records.

How far back the Astrological Aztec records go has not been able to be determined because of the invasion of Mexico by the Spaniards and their attempt to destroy all records. All the paintings and records were collected, piled up and burned. So who knows how far back their time tracking skills and calendars were used, possibly thousands of years. One of the calendars not destroyed was the Calendar Stone, as it was much larger, approximately 11 feet 8 inches in diameter with a weight of 24 tons and made of basaltic porphyry. After most who knew this sacred and astrological knowledge were executed, around 1551 to 1559, Friar Alonzo de Montufar had the Stone secretly buried. For over 200 years no one suspected it's existence until some workers excavating in Plaza Mayor in 1790, unearthed this magnificent testimony on astrological knowledge that now sits in the Mexican National Museum. The actual use of the Calendar Stone is uncertain, but it does contain the 20 symbols/glyphs, just as in the Mayan Calendar. Elbert Benjamine believes it was used, along with 2 other calendars to track celestial events and time.

Other Aztec Calendars

Again while much of the knowledge and calendars were destroyed, a man by the name of Diego Duran included in his work History of the Indians of New Spain (written pre 1588) he included a sketch of a swastika wheel which had the capacity for determining heavenly positions for any date. For doing this he was excommunicated and burned alive by the Spaniards. While most associate the Swastika with Germany in WW2, it has a deeper astronomical meaning.

The Swastika has 4 arms, which according to Benjamine represent the 4 Seasons and each season having 91 days. The 4 Seasons would then comprise a 364 day year. The Aztecs acknowledged the 365 1/4 day year, so after 364 days they would have 1 festival day on which no work was done and every fourth year hold 2 festival days. (It is believed the ancient Chaldeans viewed and used the Calendar in the same manner in relation to the amount of days and festivals days.)

The Star/Time Code also uses a 364 day Calendar, but instead of using festival days, for time tracking purposes, as explained in the first Chapter, it splits the extra days (12/31 and 2/29) between the days preceding and following the date. Many still happily use 12/31 as a festival day, but it's doubtful they know of its origin and purpose. While the Aztec Swastika Calendar reveals a 364 day Calendar, it is still common belief the Aztecs used a 360 day one with 5 extra days.

20 Day Increments or Something Different?

We know that the accepted Mayan information tells us there are two 20 day/month calendars, the thirteen month (20 days per month) calendar with a total of 260 days and the eighteen month (20 days per month) calendar with a total of 360 days. The theory on the Aztec Calendars is pretty much the same.

Benjamine continues in the article to describe the concept of the widely accepted 20 day increment of months is mistaken and that historians non-belief in astrological principles and symbolism is the probable cause for that mistake. He credits Edward Butt, an engineer from Kansas and president of the Kansas City Science Club for correctly deciphering the 20 pictures not as 20 days of the month, but as the names of the 13 days of the week and the names of the 7 weeks of the 91 day Seasonal Swastika Calendar.

All increments of Time in the Star/Time Code and Ancient Book of Time system are divided into 7 Planetary Periods and the 91 day Seasonal Calendar that was discovered in 1996 within the Ancient Book of Time has exactly 7 Planetary Periods of 13 days, just like the Seasonal Aztec 91 day Swastika Calendar. This just validated the discovery, or re-discovery of that cycle and place in time.

Finally, in the article Benjamine points out how the Aztecs would use three calendars: the Calendar Stone, the Triskelion Calendar and the Seasonal Swastika Calendar to track the precise location of the planets and their relationship to each other for any time, past, present and future. He also mentions some calendars had dots on them for numeric reference, but the advanced calendars and Aztec Astrologers only had pictures as they knew intimately the number of dots of each picture's value (could these actually be the cards in some form?). In the second article Elbert Benjamine goes into much more detail into the

Swastika Calendar and how it is used and the very important relationship between the Sun and the Moon.

Interestingly, the last calendar Carlos Barrios revealed is The Book of Destiny which he says is preserved in absolute secrecy by the guardians of the tradition and based on the meanings of the 20 nawales (days), the year bearer and the Mayan Cross – three calendars? Could these be the same three calendars that the Aztecs used to track the movements of the heavens above so precisely? With such similarities between the cultures one would have to come to the conclusion they do, especially with the reference to the Mayan cross and Aztec Swastika – both are crosses. The difference is, the Aztec calendars have been discovered , the pieces of the puzzle put together and interpreted correctly, (although not yet recognized by the main stream.) Some of the important Mayan Calendars are being kept secret and we aren't sure if they are being understood correctly, even by the Mayan elders themselves.

Where are the Mayans on this?

The depth and detail of Aztec Calendars and Astrological knowledge as revealed by Elbert Benjamine, seems to be far deeper than any Mayan information revealed in information available today. Just their Calendars alone are aligned with more natural daily and yearly influences as experienced here on Earth, instead of the unusual 260 day calendar, 400 day calendar and the slightly off 360 day calendar. The Aztec Calendars are constantly working precisely and perfectly together like gears in a clock just as in the Star/Time Code and within The God Clock, where as the Mayan Calendars only click together at some point in time hundreds or thousands of years later. As talked about in the movie "pi" (released in 1998), even random numbers at some point will click together.

It is most likely too much of the Mayan knowledge and wisdom was destroyed and what was left is just being used and translated in the best way people know how, but because so much is missing, the interpretation is not accurate, or just being misread. The two cultures and Calendars being so similar it could be considered the Mayans possessed and used the same information the Aztecs had, but because so much is missing from the Mayan side we are unable to see and understand it, in the way it is supposed to. The same can be found to be true with the earlier versions of the Ancient Symbolic Astronomical System (some are still in use just like the Mayan Calendar) which is known as the Ancient Book of Time – there was just too much information and detail missing and while not intentional the people using the system were just interpreting events as best they could with the limited information and understanding available.

Even with the information given to us by Elbert Benjamine and Edward Butt, still today the Aztec Calendars are being misinterpreted. Many talk and write of the grand and precise workings of these Ancient Calendars, but these two gentleman are the first to show in greater detail the actual workings of them and the deep knowledge the Ancients possessed regarding time, the influence of the heavens above and how they tracked it.

More to the Story

What not has been mentioned thus far is an interesting bit of information, in the book The Conquest of Mexico written by William H Prescott published in the late 1800's, he reveals that the Aztecs also had a system of 52 reeds divided in to 4 groups of 13, just as the regular deck of cards and that they were used for astronomical calculation. It would not be surprising to find that the Aztecs used their 52 reeds in the same manner the cards are used in the Ancient Book of Time, to precisely track time in The God Clock on a personal level. So besides the 3 calendars used by the Aztecs to track time, as mentioned by Elbert Benjamine, a fourth one now exists that would be used to track one's personal time and influences, in relation to the other 3. As will be shown in Chapter 4, this is a remarkable System unto itself, but again, built upon the same proportions as time in general and perfectly integrated within it.

Other Indigenous Cultures

Delving into the accepted information on the Mayan and Aztec Calendars, it is pretty hard to make any practical sense of it; however, while this is going on, we have had other indigenous groups that simply and naturally tracked time. One such culture was a group of Native Americans believed to be the predecessors of the Pueblo Indians that were located in Chaco Canyon in New Mexico. Some believe after the invasion of the Spaniards into South and Central America, some of the Maya migrated north into this area.

Although abandoned over 1,000 years ago, the structures built and left by these people are quite remarkable. The main central building was as big as the Roman Coliseum and a wall built in this area at Noon shows no shadow it's entire length. Other buildings were as far as 15 miles away and were in perfect alignment with the central structure area and within these structures, very detailed records were kept of the movements of the Sun and Moon.

There is an excellent documentary narrated by Robert Redford titled, "The Mystery of Chaco Canyon" that has aired on PBS and is available on DVD. One of the things shown in this documentary is a spiral sun stone set in a bluff type rock above the largest building area and how an artist one day while there, almost exactly on the Summer Solstice at Noon, watched a dagger of light come through a space created by slabs of stone positioned in an opening, point directly to the center of the stone. This discovery led to the research and realization of the importance of this area, the areas surrounding it and its mysterious history.

It is another amazing piece of history showing the knowledge and wisdom of an ancient people tracking Time and the motion of the heavens above with precision and reverence, but at the same time we are told the Mayans used calendars that seemingly go through time in their own, unusual manner, not in sync with natural cycles that surround us – again – while most other indigenous people, including the Aztecs (with the corrected information) and their calendars did.

Now What?

Looking at and knowing the Star/Time Code, the new understanding of the Aztec Calendar its workings and the remnants of the Mayan Calendar, the case could easily be made they knew and understood time much more than we realize. The mechanics and pieces of the puzzle have been there all along, we just have not been looking at it correctly, including the Mayan Elders.

All time, even the Mayan and Aztec Calendars are based in the Telios Proportions and they represent exactly what has been discussed in this book – a precise unfoldment in time – circles of time, all working together and representing the natural cycles and seasons experienced here on Earth. And that both Calendars unfold precisely and in the same sequence and order as laid out in the first chapter – the 364 year Cycle, the 91 year Cycle, the 28 Saturn year Cycle and the 7 year Cycle. Then we have the same Code unfolding in days – 364 day Solar Cycle, the 91 day Seasonal Cycle, the 28 Day Lunar Cycle and the 7 day Weekly Cycle, a linear code that simply and beautifully repeats itself – 364....91....28....7....364....91....28....7.

And each cycle divided into 7 periods of time and each one of those periods assigned a planet to govern it in order from Mercury through Neptune. I do not think it is intentional, but again, with much of the information missing (destroyed by the Spaniards) and the misinterpretation of the information available, we have ended up with the current ideas on what the Mayan and Aztec Calendars are, or are not.

A Final Thought on the Mayan and Aztec Calendars

It is most likely, the symbols within these calendars are defining blocks of Time, both large and small tied together in cycles and in perfect proportion. The people and culture were different from ours, so it was probably written and illustrated in a manner they could understand, even if it was just for the high priests at the time.

Besides the missing information and possible misinterpretation of the calendars, there is another challenge, the language and its possible limitations. The original language spoken by the early Mayans is referred to as Proto-Mayan and currently there are 18 different dialects of the this being spoken today. They might look at a symbol and interpret it as some kind of vertical ascension, or rise of the spirit, where as we would look at it and think, ladder. Again, are we looking at it correctly?

We will get into much more detail in Chapter 4, of the structure of the Ancient Book of Time (the Aztec System of 52 Reeds) along with all the Cycles and Planetary Periods with their influence, which represents the natural cycles and seasons we experience here on Earth in precise cyclic time. Then, as mentioned, in Chapter 6 we will get into much more detail in the layout of time in what we will call The Mayan Time Grids. One could make the case from the remnants of their once great civilizations and calendars, that at one time it was also known and used by both the Mayans and Aztecs with great understanding and precision.

Even though these calendars are being interpreted incorrectly, just as with a broken watch, it is still accurate twice a day and at some point, something in these calendars will seem to hit, but will just be used as a justification to keep interpreting them in an incorrect manner.

The end of the Calendar Chapter.....

Chapter Three

The Great Pyramid of Egypt and
the Pyramid of the Mayans

"not a stone has been set, not a dimension has been determined, which may not have its reason why, and concurred to establish scientific formulae to represent, and to eternally preserve, the previous knowledge acquired by the immortal architects of this colossal edifice." From the great French savant, M. Dufeu on the Great Pyramid

Of all the ancient structures in the world, two of them stand out in regards to time– Cheops, the Great Pyramid of the Giza Plataea in Egypt and El Castillo, the Mayan Pyramid in Chichen Itza Mexico. Both have been studied extensively and much has been written on what they possibly signify, however, knowing the Star/Time Code and the Ancient Book of Time, looking at these structures differently, it may reveal the real message that was intentionally hidden within these ancient architectural wonders. And, could either of these pyramids represent The God Clock in stone?

The Great Pyramid of Egypt
General Information

Located on the Giza Plataea in Egypt, the Great Pyramid is the only surviving structure of the 7 Wonders of the Ancient World. The accepted explanation is that it was built approximately 4500 years ago (2580 - 2560B.C.) and that it was built by the Pharaoh Khufu as either a monument to himself and/or as a tomb. It rests on approximately 13 acres, had an original height of 485 feet, has 4 sides: each side a triangle with its original base length of 9131 inches and the total original perimeter base measurement of 36524 inches. The angle, or inclination of the sides is 51.51 degrees and it was made with 2.3 million stones that weigh anywhere from 2.5 tons up to as much as 15 tons.

It is believed the builders knew of "pi" as J. Ralston Skinner eludes to in his Source of Measures referring to the Great Pyramid – "Another may be taken as geometrical, as it was so built that its height should be to one-half its circumference, as diameter to circumference of a circle. Hence it exhibits itself as one not only monumenting a method of quadrature, the elements of which we posses, but also a measure of the sun's time and also the inch and foot values."

There are also the chambers, halls and pathways within the pyramid, which we will not get into in any great detail in this work. Dr. Brown Landone did state that the King's Chamber was one of the 7 Sacred Temples in the world and that it and the pyramid were built to the Telios (perfect) Proportions.

The Deck of Playing Cards
and the Great Pyramid

In the book Nuggets From King Solomon's Mine by John Barnes Schmalz published in 1908 the author gives a detailed illustration of the pack of 52 playing cards and reveals correlations between the actual size of the deck of cards, astronomical symbolism and measurements and how they are encoded in the Cards and in the pyramid, as well. In Chapter 2 he states – "Man has thus been the agent, psychologically guided, to preserve and perpetuate and to finally assemble into a composite unit the record of the Great Pyramid. Only in this latter day, since the nature truths so marvelously depicted in the Great Pyramid have been revealed through modern scientific investigations, could this record been found in the cards and interpreted."

Another very interesting correlation between the Playing Cards and the Great Pyramid is done by Willis F Whitehead in his book The Mystic Thesaurus published in 1899. In it he states "the cards also posses a definite time value. The pyramid incline 10 to a rise of 9 feet is a key to this time value. Dividing 10 by 9 gives the time value of the Ace of Hearts – 1 minute, 6 seconds and 40 thirds. The time value of the 2 of Hearts is twice this, or 2 minutes, 13 seconds and 20 thirds. The value of the 52nd card , the King of Spades, is 52 times this unit of 1 minute, 6 seconds and 40 thirds. Now subtracting the time value of the Aces, we have in the sum of the remaining 48 cards, the exact day of 24 hours."

Whitehead doesn't explain why the Aces are taken out, but they could possibly represent the four zero cardinal points in the Zodiac and therefore removed.

These interesting facts on the playing cards and the Great Pyramid are mentioned as it clearly reveals the math, symbolism and astronomical information that the Ancient Book of Time is based on, is deeply encoded within this grand structure.

Is Something Off?

It is accepted that the angle that would be formed at the top of each side of the pyramid if it came to a point (the cap stone is missing) is 61 degrees. But, what if the angle is really 60 degrees, after all the structure itself is thousands of years old, has probably settled slightly and is now the angle at the top 61 degrees? The inclination, or angle on the side is presently measured at 51.51 degrees, so if it has settled, one has to wonder, was the original angle of the side when they were built exactly 52 degrees, less than a tenth of a degree difference? There are 52 years in the 7 planetary periods in the corrected Mayan 364 year cycle, 52 weeks in a year, 52 days in the 7 planetary period of the year and 52 cards or symbols in the deck of 52 playing cards – coincidence?

More to the Angles

Building a pyramid out of let's say 3/4"wood, one would have to set their saw at specific angles to get precise cuts for it to fit together perfectly. First, to create an equilateral triangle, where each angle is 60 degrees, a saw has to be set at 30 degrees to make each cut. In order for the 4 triangles to fit together, when making the 30 degree cuts, it will also have to be beveled 7 degrees, which we will see in a bit is also an important Number. After fitting each piece together to form the pyramid, comparing this to a pyramid with 61 degrees at the top angle of each triangle to the naked eye, little, if any difference can be seen. Changing the top angle of each triangle, or side from 61 degrees to 60 degrees would raise the pyramid ever so slightly and would bring the inclination, or angle of the side closer to 52 degrees, if not exactly that. We have to ponder then, is the top angle of each side slightly off intentionally, from settling, or when the outer casing was still intact was the pyramid made from 4 equilateral triangles? Either way, the angles are so close and other measurements so remarkable that the message of the builders is very clear.

Astronomical Angles

Within all the angles in and around the Great Pyramid lies the important astronomical aspects that are formed by transiting planets, both between the planets themselves and to those in one's Natal Chart (A chart cast at the time and place a person is born to reveal a picture of the heavens above at that moment.) Remember the saw had to be set at 30 degrees to make the cuts to form a 60 degree angle. 30 degrees in astronomical terms is the amount of degrees in each of the 12 Zodiac Signs – 12 times 30 equals 360, the number of degrees in the circle.

Each triangle, or side of the Pyramid has three 60 degree corners, 60 degrees represents what is called a Sextile. This is considered a fortunate angle or aspect. Two corners of the triangle would be 120 degrees, this is known as a Trine – another fortunate aspect. Three corners would form 180 degrees, this is known as an Opposition and when this aspect is present, friction, or difficulty is created (think of 2 magnets pushing away from each other.) The base of the pyramid has 4 corners of 90 degrees and 90 degrees forms an aspect known as a Square. Squares are considered another, not so nice aspect.

All of these aspects are experienced at some point in our lives – they can set up certain gifts or lessons to over come in our Natal Chart. We will experience them as transiting Planets from aspects or angles to the planets in our Natal Charts creating the environment in which significant events or periods in our lives will transpire in our lives. There is also the aspects or angles formed between the transiting planets themselves that influence Earth and us as a whole.

Symbolic/Spiritual Angles

As shown, the regular deck of playing cards is also represented in the Great Pyramid in remarkable ways. We have shown that the angles formed within each component, or piece of the pyramid (the four sides and the base) illustrate a significant purpose, but what about the angles created and formed when the pyramid is put together?

As mentioned, bevels of 7 degrees have to be cut on each side of each triangle in order for them to fit together. Also, the inclination, or angle on each side is, or almost exactly 52 degrees, 52 reducing to the Number 7, (5 + 2 = 7). In Ancient Hebrew/Chaldean Number System the Number 7 is associated with the Planet Neptune, the Planet that governs the Spirit/Spirituality. Before the re-discovery of Neptune (in 1846 by astronomer Johann Gottfried Galle), the Number 7 was considered the secondary Number of the Moon. At some point after the discovery, the Number 7 was rightly given to the association with Neptune. Just as with astronomical rulers and governance associated with specific planets, numbers, through their nature and qualities (along with possible observation over long periods of time) are also given associations with specific planets. The Number 7 again, represents things Spiritual.

Besides Willis F Whitehead's incredible revelation on the 24 hour day time value derived from the pyramid's inclination, the 52 degrees also represents the 52 Symbols within the regular deck of playing cards and the Aztec

System of 52 reeds. Again the regular deck of playing cards representing The Ancient Book of Time, an old Symbolic Astronomical System

Time, the Environment and the Path

So standing out in the open, right in front of us, hidden within this remarkable ancient structure are three important things. The first is Time, shown in the actual measurements themselves. The 9131 inches (91.31) on each side representing the number of days in each Season and the 4 Seasons themselves, the total base measurement of 36424 inches (364.25), the number of days in 1 Year and if a circle is placed within the square base of the Pyramid it would have a circumference of close to 28671 inches (28.671) the approximate years of Saturn's orbit and the number of days of a Lunar orbit. Along with that the 24 hour day derived from the inclination, or angle of the side and the division of Time itself by 7 revealed in all angles that are 7 degrees or reduce to the Number 7. And the Number of Saturn (8) the planet that governs Time, hidden within the original height, 485 feet (4 + 8 + 5 = 17, 1 + 7 = 8.)

The second is the Environment, in all the astronomical information revealed in all the angles formed by and in building a pyramid – the 12 Signs of the Zodiac through the 30 degree angles, all the major aspects experienced by the transiting planets in the main angles in the pyramid, the 7 major planets (Mercury through Neptune) in the Number 7 angles (the side inclination and the bevel formed between each side) and the Saturn/Lunar/Moon increment as shown in a circle drawn within its base.

The third is the Path shown in the Symbolic Astronomical System (the Ancient Book of Time) hidden within the regular deck of 52 playing cards, both by revealing the 52 symbols themselves through the 52 degrees of the inclination or angle of each side and using them to derive a 24 hour day and the four 0 degree Cardinal Points in the Zodiac represented by the Aces being pulled out of the deck in the 24 hour day calculation.

The Journey of the Soul

These three things within the Great Pyramid are truly remarkable and may point to the real message that the builders left for us to figure out. When we are born here on Earth in physical form in a specific place at a certain moment (in time), we are on a precise PATH as defined by our place within and by the 52 Symbols, an ancient science hidden within the regular deck of playing cards. This PATH will take place within the Astronomical blueprint defined the moment we are born and the motions and positions of the heavens above during each lifetime creating the ENVIRONMENT in which this will happen. And all of this happening not only within TIME, but THROUGH Time as shown in our journey upward, back to the center, or source by the height (485/8 Saturn) of the pyramid.

We will experience the PATH of the 52 Symbols by our own PATH that we will go through for good or for bad. We will go through the meaning of each Symbol through our being that Symbol, through the events and experiences associated with it and through/with other people, as well. While we are on this PATH different areas will come to light for good or for bad as defined by the ENVIRONMENT created by the Astronomical influences present at the beginning of our path and along the way – again this all happening within and through TIME.

The angle of the sides is 52 degrees, which reduces to the Number 7, the number of Spirit. So along with the PATH, the ENVIRONMENT and TIME, this inclination or, angle might represent the upward spiritual (sometimes difficult) journey of the soul through and within these three influences back to the Source (God?)

The God Clock in Stone?

Do we live more than one lifetime? Is this how the Soul evolves? Is the beginning and/or center of the God Clock the source from which all that we know comes. Is it where after taking all the PATHS possible under all the ENVIRONMENTAL influences within all the Circles of Time we end up – the beginning? Does the Great Pyramid of the Giza Plataea then represent not only the Journey of the Soul back to the source, but The God Clock itself, and how it works, or at least a portion of it as we experience it here on Earth?

The Center

So we have the possibility that the Great Pyramid represents The God Clock in stone (or a portion of it) and that the center of The God Clock may be the source from which all is created, emanates from and is the possible destination of the long journey through the Circles of Time each Soul takes. This is interesting as the Great Pyramid is located exactly in the center of Earth's land mass – its east/west axis passes through Africa, Asia, and America, the longest land latitude on Earth and it north/south axis passing through Asia, Africa, Europa, and Antarctica, the longest land longitude.Another interesting point regarding the center and the Ancient Book of Time, is the Plate known as the Spirit Plate (God's Plan) where all the cards are laid out in the 7 columns and 7 rows, with the last 3 placed on top in perfect order from the Ace of Hearts to the King of Spades, the Queen of Clubs sits in the center of the Plate. Where does the Queen's Chamber lie within the Great Pyramid? – exactly in the center.

Mystic ♠ ♥ S ♦ ♣ Quadrate

			K♠ K♠ K♠	Q♠ Q♠ 8♦	J♠ J♠ 10♣			
☿	7♡ 7♥ A♠	6♡ 6♥ 3♦	5♡ 5♥ 5♣	4♡ 4♥ 10♠	3♡ 3♥ Q♣	2♡ 2♥ A♣	A♡ A♥ 3♥	☿
♀	A♣ A♣ 2♥	K♡ K♥ 9♠	Q♡ Q♥ 9♣	J♡ J♥ J♥	10♡ 10♥ 5♠	9♡ 9♥ 7♦	8♡ 8♥ 7♥	♀
♂	8♣ 8♣ 8♣	7♣ 7♣ J♠	6♣ 6♣ 2♦	5♣ 5♣ 4♣	4♣ 4♣ 6♥	3♣ 3♣ K♦	2♣ 2♣ K♥	♂
♃	2♢ 2♦ A♦	A♢ A♦ A♥	K♣ K♣ 8♠	Q♣ Q♣ 10♦	J♣ J♣ 10♥	10♣ 10♣ 4♠	9♣ 9♣ 6♦	♃
♄	9♢ 9♦ 5♦	8♢ 8♦ 7♣	7♢ 7♦ 9♥	6♢ 6♦ 3♠	5♢ 5♦ 3♣	4♢ 4♦ 5♥	3♢ 3♦ Q♦	♄
♅	3♠ 3♠ J♦	2♠ 2♠ K♣	A♠ A♠ 2♣	K♢ K♦ 7♠	Q♢ Q♦ 9♦	J♢ J♦ J♣	10♢ 10♦ Q♠	♅
♆	10♠ 10♠ Q♥	9♠ 9♠ 6♠	8♠ 8♠ 6♣	7♠ 7♠ 8♥	6♠ 6♠ 2♠	5♠ 5♠ 4♦	4♠ 4♠ 4♥	♆
	♆	♅	♄	♃	♂	♀	☿	

Other Theories

Besides the standard, accepted explanation of the Great Pyramid of who, how, why and when the pyramid was built, others have not only questioned this, but suggested newer, actually older theories. Graham Hancock author of Finger Prints of the Gods and Robert Buvall/Adrian Gilbert authors of The Orion Mystery have not only proposed that the Pyramids/Giza Plataea were built much earlier but actually mirror star alignments in the sky – Orion's Belt. Through the use of computers, the time at which the layout of the Giza Plataea would match the sky above would be around 10,500 BC. Robert Buvall has also proposed that the shafts from the chambers within the Great Pyramid extended out into the heavens and points to specific stars that may have been important to the Ancient Egyptians and their culture.

Through the research presented by these and other authors, we are being shown the possibility the Great Pyramid might be much older than previously thought, but more importantly it may reveal a profound connection that we have here on Earth to the heavens, or even somewhere out in space. As The God Clock illustrates, we are indeed connected to other things out there more than just here on Earth and the astronomical position/s pointed to by the Great Pyramid has possibly revealed where our Circles of Time have connected not only to another time and place, but to possible beings from somewhere out in the Galaxy, as well.

So who Really Built the Great Pyramid?

From Dr Browne Landone's book Prophecies of the Melchi Zedek in the Great Pyramid and the Seven Temples – "These facts indicate that there has been and is a great Order of Mystics who – for thousands of years – have possessed a remarkable knowledge of astronomy and mathematics and building – as well as knowledge of the nature of man and the time-rate of his progress on earth – and that the mystic wise men of this order built the facts they knew into the structure of each key temple of each great civilization; and that they designed the Pyramid of Gizeh also."

"The first evidence of this is the fact that the only recorded data of the design of the Pyramid are not found in Egypt or Africa, but in treasured scraps of very old manuscripts, in the sacred libraries of the Masters of Tibet."

So were they built around 10,500 BC, or at least much further back in history than the accepted theory of 2500 BC as some propose, or not? Dr Brown Landone suggests the Great Pyramid was possibly planned by holy advisors to Khufu as not only a monument to him, but to be a temple of temples, which by its structure would embody records of the knowledge and prophecies of all key temples on earth and was built around 2881 BC.

And that even though the plans for the pyramid were with Masters of Tibet, the actual origin of them were from the Eastern Andes and that it may have been the mystics of the Andes who actually built the Great Pyramid of Gizeh.

Not Who or When but What

There are many theories as to who and when the Great Pyramid was built and at some point it will most likely be discovered and be remarkable, but now, more importantly is what it means, or what it stands for. Within all the angles, measurements and symbolism integrated into the Great Pyramid is hidden the message the builders left us – the journey of the soul on a specific path, within an astronomical environment, in and through the circles of time, back to the source or center – The God Clock. And looking at it differently through the stars, mathematics and symbols, it is so obviously clear.

El Castillo
The Mayan Pyramid

The Mayan Pyramid is in Chichen Itza, Mexico and although not as seemingly impressive as the Great Pyramid of Giza, it is still quite remarkable in its own right and its builders may have encoded important Time and Path information within it, as well. Down through time we have tried to understand its message and purpose and by using the Ancient Book of Time and the language of Time, we just might decode it.

General Information

The pyramid in Chichen Itza known as "El Castillio" was built between 800 to 900AD (El Castillio in Spanish means "The Castle".) It is also known as the Temple of Kukulkan and it is believed this pyramid was built to honor Kukulkan, the Winged Serpent. The story goes Kukulkan came down from the heavens and taught the Mayans all about mathematics, astronomy and the arts and then returned to the heavens above. It is close to 98 feet tall, each of the 4 sides are approximately 182 feet long, has a stair case with 91 steps, as well as, 52 panels and was built over an older, smaller structure. Its claim to fame is on both the Spring and Fall Equinoxes a chain of shadow and light is formed on the sides of the stair case representing a serpent (Kukulkan) descending from and ascending to the heavens above. It is positioned east of True North 19 degrees in order for this phenomenon to happen.

The Great Pyramid vs The Mayan Pyramid

There are some interesting comparisons between the Great Pyramid of Giza and the Mayan Pyramid of Chichen Itza. At first glance they look very different – the Great Pyramid is a more classic pyramid built upon 4 equilateral triangles (or very close to being equilateral) and the Mayan Pyramid is more of a step pyramid with its 9 levels and its actual 91 steps on each side. While the angle of the Great Pyramid is approximately 52 degrees, the Mayan Pyramid's angle is 45 degrees, but has 52 panels on each side.

The measurement of the base of each of the 4 sides of the Great Pyramid is 9131 inches referring to the 91.31 actual days in a Season, the Mayan Pyramid has 91 steps on each side referring to 91 days in a Season. This is interesting as in Chapter 1 we mentioned Linear/Sequential Time vs Cyclic Time – the Great Pyramid in its measurements is possibly representing Linear/Sequential Time and the Mayan Pyramid in its 91 steps is representing Cyclic Time. When the lengths of the 4 sides of the Great Pyramid are added together we get 36524, or 365.24, the number of days in a year. When we add the 91 steps of the Mayan Pyramid together we get 364, the number of days in a year within the Star/Time Code. The Temple on top of the Mayan Pyramid would then represent the last 1.24 days. As we know, our calendars do not accurately reflect linear time, therefore an adjustment needs to be made every 4 years to bring it back in sync. It is possible the Mayans knew of both Linear and Cyclic Time and it is reflected in their pyramid, El Castillo. Also, the estimated original height of the Great Pyramids is 485 feet and the approximate height of the Mayan Pyramid is 98 feet – when each of the integers of each height is added together (4 + 8 + 5 = 17, 1 + 7 = 8) and (9 + 8 = 17, 1 + 7 = 8) it reduces to 8, the Number of Saturn which is the Planet that is given the governance of Time.

The Ancient Book of Time
the Mayan Pyramid and the Path

As we have shown there are some remarkable correlations relating to the regular deck of playing cards and the Great Pyramid, while much of that information is apparently hidden, they aren't even trying to hide it in the Mayan Pyramid. In a Deck of Cards there are 4 suits and when the spot values of each of the card in a suit are added up from 1 through 13, the total is 91, the number of steps on each of the 4 sides of the Mayan Pyramid and the number of days in each Season. Add the 4 sides together, just like in the 4 suits in a deck of cards we get 364, the number of days in a year in the Star/Time Code. In the Deck we add the Joker for the last 1.24 days and with the Mayan Pyramid, the Temple on top is added for the same purpose. As will be shown in the next Chapter, the Joker mathematically may actually represent life manifest on the material plane, The temple on top of the Mayan Pyramid then, maybe honoring this same miraculous possibility.

There are 52 cards in the deck and there are 52 panels on each side of the pyramid. There are 4 suits within the deck of cards representing not only the 4 seasons, Spring, Summer, Fall and Winter, but the 4 elements, Fire, Air, Earth and Water. This is perfectly reflected in the Mayan Pyramid and could very well point to the fact that the Mayans understood the details of the Path as revealed in the Ancient Book of Time.

The Environment
Where is it?

The Astronomical information is encoded into the Great Pyramid in its angles and shape, but where is it in the Mayan Pyramid? The Mayan Pyramid itself has 9 steps, or levels and what could these represent? When we go through our environment in an astronomical sense, there are 9 main influences we experience, the Sun, Moon and the 7 planets, Mercury through Neptune, this gives us a total of 9. Could this then, be the real message of the 9 levels built into the Mayan Pyramid? Are they also trying to tell us that not only are we here for a specific path in this life and that we will experience it in an astronomical environment? And since each of the 4 stair cases are surrounded by all 9 levels and if we didn't figure out what the significance 9 planetary influences were, did they create an environment for dummies to literally show us that our Path is again, surrounded within this astronomical environment?

And of Course...Time

While there doesn't seem to be any clear reference to Time in the Mayan Pyramid, although it is a closer reflection to our Calendar than most of the other accepted Mayan Yearly Calendars, there are a couple of hints within it. When the phenomenon of the chain of shadow and light are revealed on the side of the stair case on both the Spring and Fall Equinoxes, 7 triangles appear at that time. This could possibly represent not only Kukulkan descending and ascending from/to the heavens above, but time itself being divided by Spirit, or the Number 7. Remember each time cycle within the Star/Time Code is divided into 7 Planetary Periods. Also, the height of the Mayan Pyramid is approximately 98 feet and the total length of all 4 sides of its base added together is 728 feet, when we divide the total base length of 728 by its height, 98 feet we get a rounded off sum of 7.428571. 7 again the Number of Spirit and that which divides or runs through all and the remainder 428571 is actually the numbers within the Sacred Number 142857 which reduces to the Number 9, the true Number of man and the manifestation of life on the material plane. The Sacred Number (142857) is arrived at through the simple remainder of dividing a circle's circumference by its diameter and will be dealt with in more detail in the next chapter. Could it be they are trying to illustrate to us some important correlation or relationship between Spirit (7) and Man (9) and/or Life here on Earth – the spiritual journey of the Soul? Finally, as mentioned, the height of 98 feet adding to the Number 8 (9 + 8 = 17. 1 + 7 = 8), the Number of Saturn, the planet that rules, or governs time.

Where are Mayan Calendars?

In Chapter 2 we looked at and questioned the timing and basis of the accepted Mayan Calendars and a possible alternative, or correction to them. Those calendars were the 260 day, 360 day and 400 day, with the 260 day and 360 day calendars being based on their respective 13 and 18, 20 day months. The question has to be asked, how are these calendars reflected within the Mayan Pyramid? Some say the 9 levels, or steps are divided by the stairs, giving us 18 levels to represent the 18 months in the 360 day calendars, the problem is there are still 9 levels and there is no mention of the 20 days, or the 13 months in the 260 day calendar. Besides this, there really isn't any correlation to any of the other calendars mentioned within, or on the pyramid. If these Calendars were so important, including the larger one based on years, why weren't they reflected in this auspicious structure?

The 52 panels on each side of the Mayan Pyramid are said to represent some larger 52 year cycle. We know from Chapter 1 that 52 as an increment of time is not a cycle in and of itself, but one of 7 planetary periods within a cycle – be it a 52 day planetary period in a yearly cycle or a 52 year planetary period in a 364 year cycle. The 52 panels may actually on some level, point to the time increments within the Star/Time Code in both days, years, or even larger blocks of time.

The Message of the Mayan Pyramid

As we know, each of the 4 sides of the Great Pyramid are 52 degrees and each of the 4 sides of the Mayan Pyramid has 52 panels, which may actually represent the 52 symbols in the Ancient Book of Time and that they/we need to experience each of the 4 Seasons and/or elements through these symbols in order for the soul to fully evolve in its journey back to the center, or source.

Another interesting point is the length of one side (approximately 182 feet) of the Mayan Pyramid would be the diameter of a circle that would fit within the square of the base of the pyramid. When 182 is multiplied by "pi", or 3.14, we get a total of 571.48. When these integers are added together we get a total of 25/7, 7 being the Number of Spirit and Spirit runs through and divides everything, including time. As we will see in chapter 5, the Number 7 is important regarding the center and it could be the Mayans were trying to tell us something in relation to that. We start as Spirit, emanating from the center in possibly The God Clock, honoring our journey through time.

Just as in the Great Pyramid, the message of the builders of the Mayan Pyramid encoded within it the journey of the soul through an astronomical environment, along a specific path, within and through time, in each life, or lifetime – they just did it in a different way. Some of it strikingly obvious and other parts in a not so clear way. We know much is missing from the Mayan information, but we do know the Aztecs had a system of 52 reeds used for astronomical calculations and that it was most likely used in the same way as The Ancient Book of Time (an old symbolic astronomical system based on the math and symbology behind the regular deck of cards) – as illustrated by El Castillo, the Mayan Pyramid in Chichen Itza, Mexico, the Mayans must have known of and used it, as well.

One has to wonder that if Dr Browne Landone was correct in saying the mystics from the Eastern Andes had something to do with the building of the Great Pyramid, and if so, with Chichen Itza being much closer, that they to, may have hand in the building of the Mayan Pyramid, as well. But with the cultures being so different, the language, or code in which it was built was different, but revealing the very same message.

The end or last stone in the Pyramid Chapter….

Chapter Four

The Ancient Book of Time
The Amazing Deck of Playing Cards

When we think of a deck of ordinary playing cards what comes to mind? The usual card games such as Poker, Black Jack, Bridge, Solitaire and others. Many talented people can do incredible tricks with playing cards, and, of course, an old grandmother or aunt who can read fortunes using a deck of cards. While all of this may be amusing and true, there is a much deeper meaning and essence behind them.

A Little History

Playing Cards are believed to have originated in China in the 12th century where paper was first invented. Earlier than this, they were believed to be in Persia and the surrounding region, but without paper were they woven, painted on wood, or some other material? It is thought the Gypsies may have brought them to Europe and in the 18th century a book was published titled The Square of Sevens by Robert Antrobus, revealing an actual method of Gypsy card reading. There is even a Gypsy Witch Deck available today designed specifically for card/fortune reading

Another possibility given is that military servants in Egypt used a deck of 52 cards, very similar to the Italian decks and that they came to Europe by crossing the Mediterranean in the 14th century. From there in a short time they were used all over Europe and as printing capabilities improved, so did the decks. Within this time frame Germany made a giant leap forward in printing and were turning out decks by the thousands.

The deck went through many changes – the Court Cards changed, both in style and layout, the number of cards in the deck, some with 48 cards, some with 56, but eventually evolved in what we have today, by which it is probably no accident. The Gypsy's "Square of Sevens" (as you will see) is interesting from a mathematical layout point of view and the Egyptian military servants using a deck of 52 cards, as well, since this information is encoded into the Great Pyramid.

One has to wonder, as Dr. Brown Landone suggests, that if all the sacred knowledge originated in the Eastern Andes and then somehow, it was spread out into specific places in the world in some manner, did the Mystics from the Andes bring with them, not necessarily the actual deck of cards, but the symbols/system in whatever form it was in just to get it out there in some way? After all, the Aztecs had their system of 52 reeds used for astronomical calculations and they surely did not get that from the Spanish Conquistadores, as they were trying to destroy all the Mayan and Aztec information. The Aztecs were not only using the 52 Reed System in a much more mathematical and purposeful way than the more civilized societies in Europe, but well before the cards even showed up in Europe.

The More Modern History

Then in 1893, Olney Richmond, the self proclaimed Grand Master of the Order of the Magi published a book titled The Mystic Test Book and within it revealed some of the deeper mysteries behind the ordinary deck of playing cards. Richmond had an Order of the Magi Temple in the Chicago area where he taught some of this information and promoted the Religion of the Stars. Willis Whitehead, an author mentioned in the last chapter was also in the Chicago area at that time and attended some of Richmond's classes. Whitehead himself was in a group called The Star of the Magi and had several accurate, predictive astrological articles published in their newsletters that ran from November of 1899 to 1902. He said that Richmond claimed to be a Professor of Astrology, but actually quit practicing it as he wasn't very good at interpreting astrological charts. According to Whitehead, Richmond also believed he was the reincarnation of Jesus Christ.

Just as Elbert Benjamine mentioned in the second chapter, the same problems with the academics misinterpreting the Mayan and Aztec Calendars because of their lack of astrological understanding, the same thing happened with the Card System. Richmond was supposedly handed this information by some mysterious man around the mid 1800's and while good at mathematics, his lack of understanding regarding the unfoldment time and the Star/Time Code most likely (just as in the Mayan and

Aztec calendars) sent the Card System in a wrong, slightly off direction.

In 1947, a book on the Card System was published by Edith L. Randall and Florence Evylinn Campbell titled Sacred Symbols of the Ancients. Finally, this book revealed some of the partial workings of the System. It gave a description of each card as a person, similar to the interpretation of a Natal Chart in astrology and in a later published version, included what they called the 7 Year Spreads and the 90 plates, or quadrations. Unfortunately, with much information missing in interpretation and understanding, techniques were created to fill in the gaps that were neither accurate nor correct. This was not done intentionally, it was an honest attempt to more fully use the System with important pieces of the puzzle still missing.

Since then, others have written books and have taught the System, one even claiming to have been ordained the Grand Master of the Order of the Magi on the astral plane. However, no real change to the deeper workings, or understanding of it came about. The System, as being explained and taught in the 2nd half of the 20th century, seemed to be operating by its own rules and along the way with even more unintentional but misguided techniques being taught as truth, just as with the Mayan and Aztec Calendars. Luckily, some of the System was accurate enough to keep it going and the notion that at its core, it was a deep, mysterious System only known to those who have been initiated into some higher spiritual realm created interest in it.

A Mystery Revealed
The Puzzle is Solved

In the mid 1990's, using and studying what was available on the System, the Star/Time Code, a mathematical sequence of Numbers that relates closely to the orbits of Saturn, the Earth, the Moon and Seasons of those motions, was discovered. This led to a much deeper understanding of not only the Card System and the workings within it, but was the key to unlocking other mysteries, as well.

So What Really are the Cards
and How do They Work

The regular deck of playing cards is actually an ancient science, an old symbolic Astronomical System based on the mathematics and symbology within the cards themselves. And just as mentioned with the Aztec's 52 reeds, they are used for astronomical calculations. They are an exact reflection of our calendar – 52 weeks in a year, 52 cards in the deck, 4 seasons, 4 suits and when all the spot values of the cards are added together (1 through 13) in each suit, we get 91, the number of days in each season. When the 4 suits are added together, we get 364 and added to that the Joker to get 365, the number of days in a year. The 12 Court Cards represent the 12 months and the 13 cards in each suit also representing the 13, 7 day weeks in each Season and the 13, 28 day lunar cycles within each year.

Each card has a symbolic value from 1 through 52 and also a numeric value of 1 through 13 (Ace through King) and a planetary association (Sun, Moon and the 7 planets Mercury through Neptune.) The Suits themselves not only represent the 4 seasons (Hearts/Spring, Clubs/Summer, Diamonds/Fall and Spades/Winter), but also the 4 elements, as well – Hearts being Fire, Clubs Air, Diamonds Earth and Spades Water. The qualities created from the components of each card give it both a meaning and an essence. As a placement in any of the various charts, it can point to an experience, event, or even a specific person, either by its nature and/or specific card of another person.

As an actual person, a card can reveal a person's personality, nature, physical stature, mental potential, etc.

Each day of the year then, through a mathematical calculation is given a card. Some cards get more days than others depending on the number of the month and the day and what it calculates out to. For example, the only day that is given to the Ace of Hearts is December 30th and the only day given to the King of Spades is January 1st, while the Ace of Diamonds is given a total of twelve different days. The Ace of Diamonds reflects a desire for worth and material things, so in today's world it is of no surprise many dates/persons are that Card.. The card represented by the day on which a person is born is then known as the Birth, or Sun Card of that person.

Since the actual calendar of 365 days with an extra day added every 4 years and the Card System is based on the Star/Time Code Year of 364 days, two days have to be dealt with, or adjusted for the System to work. In dealing with February 29th (the extra day added every 4 years), the accepted method was to use the month/day calculation to come up with its own card, the 9 of Clubs. December 31st was considered the Joker and anyone born on that day, the System could not be used as they could be any card in the deck they wanted to be (a Joker's wild kind of deal.) First, just because you are born on December 31st does not mean you get to live your life as any card you want as you are subject to the same influences on Earth as everyone else. In dealing with February 29th, you can't have a System based on 364 days for 3 years, then every 4th year change it to a 365 day System.

To simplify it, in the Ancient Book of Time System, both days are split in half and given to the days before and after. If you were born before Noon on February 29th, you would be given the card for February 28th and if born after Noon, you were given the card for March 1st. The same with December 31st, if you were born before Noon, you were given the card for December 30th and if born after Noon, you were given the card for January 1st. Actually, by straightening out this part of the System that had been used incorrectly for so long in the 20th Century, it helped unlock the Star/Time Code as everything then, fell into perfect sync.

Quadration
Thrown to the Four Directions

A deck in perfect order from the Ace of Hearts to the King of Spades is mathematically quadrated, or as the Native Americans would say, “thrown to the 4 directions” and then laid out into 7 columns and 7 rows (the Gypsy Square of Sevens?) with the last 3 cards placed in the middle on the top. This creates the Life Plate and it is from this an overall Path of one’s life is derived, similar to a Natal Chart in Astrology.

Mystic L Quadrate

			K♠ (K♠ K♠)	8♢ (Q♠ 8♦)	10♣ (J♠ 10♣)			
☿	A♠ (7♥ A♠)	3♢ (6♥ 3♦)	5♣ (5♥ 5♣)	10♠ (4♥ 10♠)	Q♣ (3♥ Q♣)	A♣ (2♥ A♣)	3♡ (A♥ 3♥)	☿
♀	2♡ (A♣ 2♥)	9♠ (K♥ 9♠)	9♣ (Q♥ 9♣)	J♡ (J♥ J♥)	5♠ (10♥ 5♠)	7♢ (9♥ 7♦)	7♡ (8♥ 7♥)	♀
♂	8♣ (8♣ 8♣)	J♠ (7♣ J♠)	2♢ (6♣ 2♦)	4♣ (5♣ 4♣)	6♡ (4♣ 6♥)	K♢ (3♣ K♦)	K♡ (2♣ K♥)	♂
♃	A♢ (2♦ A♦)	A♡ (A♦ A♥)	8♠ (K♣ 8♠)	10♢ (Q♣ 10♦)	10♡ (J♣ 10♥)	4♠ (10♣ 4♠)	6♢ (9♣ 6♦)	♃
♄	5♢ (9♦ 5♦)	7♣ (8♦ 7♣)	9♡ (7♦ 9♥)	3♠ (6♦ 3♠)	3♣ (5♦ 3♣)	5♡ (4♦ 5♥)	Q♢ (3♦ Q♦)	♄
♅	J♢ (3♠ J♦)	K♣ (2♠ K♣)	2♣ (A♠ 2♣)	7♠ (K♦ 7♠)	9♢ (Q♦ 9♦)	J♣ (J♦ J♣)	Q♠ (10♦ Q♠)	♅
♆	Q♡ (10♠ Q♥)	6♠ (9♠ 6♠)	6♣ (8♠ 6♣)	8♡ (7♠ 8♥)	2♠ (6♠ 2♠)	4♢ (5♠ 4♦)	4♡ (4♠ 4♥)	♆
	♆	♅	♄	♃	♂	♀	☿	

The Quadrature of a Circle is an interesting question or ponderance, that mathematicians have been working on for hundreds of years. As we will see in the next chapter, it may not only be the basis for the Card System (quadrating the deck), but Time itself.

The deck is then put back into the order it was before it was laid out in the columns and rows, then quadrated again and laid out into another plate of columns and rows just as before. After 90 times, the deck falls back into perfect order. Within each of the 90 plates are 52 cards in their various, yet precisely calculated positions and then 52 different Astronomical charts can be derived from each plate using each of the 52 cards as a different starting point. The Birth or Sun Card is the first of 13 cards derived from a plate both horizontally and vertically to create, or form a chart. The first 7 cards after the Birth or Sun Card are the planetary cards in sequence from Mercury through Neptune. These are both the 7 timing periods within a chart, (be it a 52 day period in a 364 day Yearly Chart, or a 4 day period in a 28 day Lunar Chart, or a planetary period from any other chart in other time cycles) and the departments such as travel, communications, financial matters, relationships, or what is ruled or governed by that planet. Mercury, for example, would rule communications and travel. The other 5 cards beyond the 7 timing ones are overall influences that are also important positions that effect the person during the time for which the chart is cast and is in effect. Shown is a chart derived for a person for that increment of time during their life. Since everything is a circle, therefore a chart is displayed in a wheel format with the Sun Card in the center just as the Sun is at the center in our Solar System. Interesting that the plates are laid out in a linear format, while the chart is laid out in a circular one possibly denoting the relationship between Linear Time vs Cyclic Time as mentioned before.

Note to Astrologers – even though the circular charts used have 12 sections within them, they are not houses as used in an astrological chart and should not be interpreted in that manner. They are simply a planetary sequential order of the cards as derived from a specific plate for a specific card and placed in a circular format.

Albert Einstein
March 14, 1879

The Card System and the Star/Time Code

As stated earlier, up until the 1990's the System and the charts interpreted within it, were only being used accurately for 91 year, 7 year and Yearly increments. Besides knowing that a year represents one of Earth's orbits around the Sun, no explanation was given as to what the 91 year and 7 year charts represented with regards to time. Re-discovering the Star/Time Code, a mathematical sequence of Numbers that relates closely to the orbits of Saturn, the Earth, the Moon and seasons of those motions unlocked not just the basis of the progressions through the 90 plates, but 5 new progressions, as well – the 364 year, the 28 year, the 91 day, the 28 day and the 7 day progressions. Using this ancient, sacred System one would not only be able to have and interpret a chart for each 91 years, 7 years and a Year and their respective 7 planetary periods of 13 years, 1 year and 52 days, as before – you now have charts for each 364 years, 28 years, 91 days, 28 days and 7 days with their respective 7 planetary periods of 52 years, 4 years, 13 days, 4 days and a day, as well. Just think, the moment you are born these Circles of Time start unfolding and the influences are calculated to a symbol and a planet down to each single day!

Step up to the Tables

We now have 90 plates, charts to be derived and looked at from those plates and 8 different progressions to use and look at, as well. Instead of trying to progress one's life from the beginning each time, a set of tables were created to reveal the starting point in the 90 plates for each year. For example, at age 55 the starting plate for the 7 day, or Weekly Progression is Plate 70. As a person is going through each of their weeks at age 55, they will progress through the plates. Since they started on Plate 70 at the 1st week, they will be on Plate 90 the 22nd week then 23rd week they will start over on Plate 1. Looking at the tables at age 56 for the 7 day Progression they will start on Plate 32. While the tables themselves are an easy reference guide as to the starting point for the various progressions each year of our lives, much more is happening within them. They also illustrate that we are effortlessly going through time, but in a very sequential, systematic order.

Again at age 55, the 7 day Progression starts on Plate 70, when 55 is added together (5 + 5 = 10, 1 + 0 = 1) we get a value of 1 and when the value of 70 is added together (7 + 0 = 7), we get a value of 7. The Numbers 1 and 7 are both Telios Proportion Numbers, in fact in all the tables from a Year and below (91 day, 28 day and 7 day) every age that is or adds to a Telios Proportional Number, the corresponding plate is also a Telios Proportional Number. In a Seasonal Table at age 55, the corresponding plate is 40, which reduces to 4 (4 + 0 = 4), a Telios Proportional Number and in the

Lunar Table at age 55, the corresponding plate is 85 which also reduces to 4 (8 + 5 = 13. 1 + 3 = 4), a Telios Proportional Number This clearly indicates that time and the path we are on within time unfolds in proportion.

Another quite remarkable occurrence, demonstrated in the tables is that while the Yearly, 91 day Seasonal, 28 day Lunar and 7 day Weekly progressions are all going through the plates at their own , seemingly unique and separate rates, (starting over on Plate 1 after the 90th plate is reached) at age 90, they all line up on Plate 90! So after quadrating a deck 90 times it falls back into perfect order and after 90 years of life on Earth, all the progressions from the Yearly on down all line up on Plate 90. This may again prove what the Native Americans believe is true, everything is a circle – even Time. Just as mentioned in the first chapter, this demonstrates an important line up, or sync and again in dealing with Time itself within The God Clock, will it to at some point, sync?

The Four Faces of Time

Just as there are 4 Suits, 4 Elements and 4 Directions, Time as well unfolds in 4 different ways: first, in a Linear Sequence as shown by the Star/Time Code – 364 years, 91 years, 28 years, 7 years, 1 year (364 days), 91 days, 28 days, 7 days and 1 day. Second, in a Cyclic manner, or cycles within cycles – inside each 364 days (1 year) you will experience 4, 91 Day Seasons. 13, 28 day Lunar Months and 52, 7 day weeks. Also, within its larger corresponding cycle (7 days/7 years, 28 days, 28 years, etc.) there will be exactly 364 of those smaller cycles – in 7 years you will experience 364, 7 day weeks, etc. Third, as shown in the tables Time is proportional as illustrated by the Telios Proportional Numbers in the corresponding Age and Plate Numbers within the tables used in the Ancient Book of Time. And Fourth, through the 90 quadrations of the deck of cards and the 90 year sync in 4 of the progressions, Time is a circle.

Intersecting Circles of Time

In Chapter 1 using the tragic event in Ferguson, Missouri on 8/9/2014 it was mentioned we would show how the circles of time and paths for Michael Brown and Darren Wilson connected. Michael Brown supposedly stole something from a store and during a confrontation with police officer Darren Wilson was fatally shot. Michael Brown was born on 5/20/1996 and the symbol, or card calculated for 5/20 is the Queen of Clubs. He was 18 at the time of the shooting and looking at Plate 18 we see that his card (the Queen of Clubs) was displacing, or sitting in the original Life Plate position of the 5 of Diamonds, the Birth Card of Darren Wilson. The 13th card, or what is known as the Phoenix Card in his Yearly Chart is the 5 of Diamonds, again the card of Darren Wilson. The Phoenix Card represents the completion, or result of that Year, or what ever chart it is in. Unfortunately, for Michael Brown, Darren Wilson was directly involved in that completion.

Mystic 18 63 Quadrate

			K♠ (K♠ K♠)	4♣ (Q♠ 8♦)	Q♢ (J♠ 10♣)			
☿	J♠ (7♥ A♠)	2♢ (6♥ 3♦)	K♣ (5♥ 5♣)	3♣ (4♥ 10♠)	4♢ (3♥ Q♣)	A♣ 2♡ (2♥ A♣)	5♠ (A♥ 3♥)	☿
♀	2♡ A♣ (A♣ 2♥)	4♡ (K♥ 9♠)	7♠ (Q♥ 9♣)	J♡ (J♥ J♥)	9♠ (10♥ 5♠)	7♢ 9♡ (9♥ 7♦)	7♣ (8♥ 7♥)	♀
♂	8♣ (8♣ 8♣)	3♢ (7♣ J♠)	J♢ (6♣ 2♦)	8♠ (5♣ 4♣)	6♣ (4♣ 6♥)	10♢ (3♣ K♦)	4♠ (2♣ K♥)	♂
♃	J♣ (2♦ A♦)	10♡ (A♦ A♥)	6♢ (K♣ 8♠)	5♡ (Q♣ 10♦)	K♡ (J♣ 10♥)	9♢ (10♣ 4♠)	8♡ (9♣ 6♦)	♃
♄	3♡ (9♦ 5♦)	6♡ (8♦ 7♣)	9♡ 7♢ (7♦ 9♥)	7♡ (6♦ 3♠)	Q♣ (5♦ 3♣)	2♠ (4♦ 5♥)	A♢ (3♦ Q♦)	♄
♅	A♠ (3♠ J♦)	9♣ (2♠ K♣)	10♣ (A♠ 2♣)	Q♠ (K♦ 7♠)	A♡ (Q♦ 9♦)	2♣ (J♦ J♣)	5♣ (10♦ Q♠)	♅
♆	K♢ (10♠ Q♥)	10♠ (9♠ 6♠)	3♠ (8♠ 6♣)	8♢ (7♠ 8♥)	Q♡ (6♠ 2♠)	6♠ (5♠ 4♦)	5♢ (4♠ 4♥)	♆
	♆	♅	♄	♃	♂	♀	☿	

The City of Ferguson would have its circles of time and path unfolding, as well. Ferguson's birth date is 11/19/1894 making it an Ace of Clubs. In its 7 Year Chart it was in it's 1 year Venus Period on Plate 17 and had the Queen of Clubs in that position – Michael Brown's Birth Card. Each person/date that is represented by a card, will have other significator cards as defined by the Astrological Sign and Number of the day itself and the respective planets that rule, or govern those Signs or Numbers. Being a Taurus, Michael Brown's Venus Card in his Life Chart is what's known as his Planetary Sign Card and that card is the 5 of Clubs. On the day of this event, the city of Ferguson Missouri in its Weekly Chart which was on Plate 15 had the 6 of Spades (one of the traditional death cards in the system) and it was sitting in the original Spirit Plate position of the 5 of Clubs, one of Michael Brown's significator Cards. In it's Chart for the year on Plate 30, Ferguson had the 5 of Clubs as its Moon Card and it was sitting in the original Life Plate position of the 9 of Spades, not only another traditional death card in the system, but also the card for the name Darren Wilson. We will get into more detail into names and the cards in the next chapter.

Mystic 15 60 Quadrate

			K♠ (K♠ K♠)	4♢ (Q♠ 8♦)	4♣ (J♠ 10♣)			
☿	Q♠ (7♥ A♠)	K♣ (6♥ 3♦)	9♠ (5♥ 5♣)	Q♢ (4♥ 10♠)	J♣ (3♥ Q♣)	2♡ A♣ (2♥ A♣)	J♢ (A♥ 3♥)	☿
♀	A♣ 2♡ (A♣ 2♥)	J♠ (K♥ 9♠)	5♢ (Q♥ 9♣)	J♡ (J♥ J♥)	A♠ (10♥ 5♠)	9♡ 7♢ (9♥ 7♦)	10♢ (8♥ 7♥)	♀
♂	8♣ (8♣ 8♣)	5♣ (7♣ J♠)	9♣ (6♣ 2♦)	6♠ (5♣ 4♣)	2♠ (4♣ 6♥)	A♡ (3♣ K♦)	7♣ (2♣ K♥)	♂
♃	6♢ (2♦ A♦)	3♠ (A♦ A♥)	10♠ (K♣ 8♠)	10♡ (Q♣ 10♦)	7♡ (J♣ 10♥)	6♡ (10♣ 4♠)	3♣ (9♣ 6♦)	♃
♄	2♢ (9♦ 5♦)	5♡ (8♦ 7♣)	7♢ 9♡ (7♦ 9♥)	K♢ (6♦ 3♠)	A♢ (5♦ 3♣)	K♡ (4♦ 5♥)	8♠ (3♦ Q♦)	♄
♅	7♠ (3♠ J♦)	4♡ (2♠ K♣)	8♢ (A♠ 2♣)	3♡ (K♦ 7♠)	6♣ (Q♦ 9♦)	8♡ (J♦ J♣)	5♠ (10♦ Q♠)	♅
♆	9♢ (10♠ Q♥)	10♣ (9♠ 6♠)	Q♡ (8♠ 6♣)	Q♣ (7♠ 8♥)	4♠ (6♠ 2♠)	2♣ (5♠ 4♦)	3♢ (4♠ 4♥)	♆
	♆	♅	♄	♃	♂	♀	☿	

There is much more detail and layers revealed in both Michael Brown's, Darren Wilson's and the city of Ferguson, Missouri's charts within the Ancient Book of Time, from the 28 Year Charts down to the day in their Weekly Charts directly related to the event. But what it clearly shows, is that 3 different entities and their seemingly separate circles of time and paths are unfolding, (two persons and a place) and at a moment in time itself, those circles intersect for a specific event to happen, for good or bad. Who knows, in the bigger scheme, it actually may have happened for good, in order for larger circles to connect, or intersect with different circles for humanity to evolve and grow.

While the progressions and charts within this ancient system are impressive in their own right, the System can be used in other ways, as well. Knowing the Life or Natal Card charts of different people, relationship comparisons can be done, revealing the nature of a relationship along with its strengths and weaknesses and even lessons to be learned. In the next chapter along with the true and deeper meaning of the Masonic Symbol, we will look into the alphabet and its relation to both numbers and the card system, as they are all tied together, or connected.

More to the Ancient Structures

Now when we look back at the last chapter on the pyramids, it is even more impressive that the Ancient Book of Time, the symbolic astronomical system based on the regular deck of cards and the 52 reeds used by the Aztecs, was encoded both physically and mathematically into the Great Pyramid of Egypt and The Mayan Pyramid in Chichen Itza, Mexico. And along with that the possibility that the 4 sides of a pyramid not only represent the 4 suits, seasons, directions and elements, but the 4 different ways in which time unfolds – linear/sequential, cyclic/cycles, proportional and in a circle and all of it happening within and possibly defining The God Clock.

Note: This chapter is not meant to be a "how to" of the System, but just to show its workings and significance as it is obviously encoded into the ancient calendars and significant structures around the world. A complete set of plates, instructions, interpretations and tables on the correct method of using this remarkable system is available through other works by the author.

The end of the Card Chapter….

Chapter Five

The Masonic Symbol
The Alphabet and Numbers

In this chapter, we are going to explore the Masonic Symbol, the alphabet and some of the deeper meanings of numbers and how they relate to what has been shown so far in this book. As you will see, it is quite remarkable and will give us a new view as to not only their importance, but show us it is the possible mathematical foundation of time, how our paths and alphabet were created and point to the existence of The God Clock.

The Masons
A Brief History

The Masons today (also know as Freemasonry), is considered one of those semi-secret societies we really don't know exactly what goes on in their organization, but some of their charity work and symbolism is right out in plain sight. Masonry is believed to have originated in Europe in the Middle Ages and spread over to the United States. In fact, many think that the United States itself was formed and based on Masonic principles, as many of our Founding Fathers were Freemasons and as will be illustrated, it may very well have been. Some guess it is possible the Masons have been around much longer, may have originated in Egypt and were responsible for building many remarkable ancient structures such as King Solomon's Temple (built around

953 B.C.), one of the original 7 Wonders of the Ancient World. The Knights of the Templar, also from the Middle Ages in Europe, which are believed to have possession of the Ark of the Covenant (a box designed by God and built by the Israelites to hold the stone tablets on which the 10 commandments were written), may also have ties to the Masons.

The Masonic Symbol

The Masonic Symbol (the Square and the Compass) is probably the most known and visible symbol of Masonry. You not only see it on their Lodges/buildings in cities and towns across the United States and else where, but on bumpers on vehicles all over, as well. The symbol itself is simply a Square and Compass with the letter G in the middle. The common Masonry meaning of this symbol is more of a moral one – the square represents the squaring of their actions by the square of virtue and the compass, circumscribing their desires and keep their passions within due bounds towards all mankind.

While this is probably all good and true, there is a deeper, mathematical side to the story.

The Masonic Symbol
The Math and the Heavens Above

As we have been talking about circles and squares in this book, through and within the various structures and systems, what then might the Square, the Compass and the letter G in the center truly represent in the Masonic Symbol? The most common meaning attributed to the letter G in this environment is that it stands for God, which may not be far off in one sense. Using the cards/symbols within the Ancient Book of Time and the alphabet/card/letter key that we will get into shortly, the capital letter G then has a symbolic value of 33 which is the solar value given to the 7 of Diamonds (the 33rd Card in the deck.) This is quite interesting as there are 33 degrees of initiation within Masonry. The Number 7, as previously mentioned is the number of Spirit and the Suit of Diamonds represents the element of Earth. The 7 of Diamonds has a Spot value of 7 as there are 7 Diamonds, or symbols on this card. Taking the compass from the Masonic Symbol, we would draw a circle with it, but what size would the circle be? Since the letter G sits in the middle/center of the Masonic Symbol and it is represented by a card/symbol with a spot value of 7, 7 would be the measurement for the diameter of the circle. As mentioned, 7 is the number of Spirit and we know Spirit runs through and/or divides everything, including time.

So if the Masonic Symbol represents in mathematical form some mystical truth, it is of no surprise that the letter G, and the Number 7 sits within its center.

A circle with a diameter of 7 would have a circumference of 22. This specific ratio of 7/22 is known as the simple, or approximative illustration of "pi", the relationship between a circle's diameter and its circumference. So what does the Number 22 signify? It simply represents the Sun, the Earth and the Moon (3), the 7 planets, Mercury through Neptune (7) and the 12 Sign of the Zodiac (3 + 7 + 12 = 22) – the 22 building blocks of our Solar System that influence life here on and including Earth. This is a beautiful, yet simple mathematical illustration that Spirit runs through, or divides our Solar System. Quite interestingly, it is no accident that the ancient Hebrew/Chaldean alphabet contains 22 letters, as well.

We now know what the compass represents, but where does the square come in, what does it stand for? When a circle's diameter is quadrated, or in Native American terms, thrown to the 4 directions, it creates a square that perfectly envelopes the circle. This square would have a measurement of 7 on each of the 4 sides and it represents the real meaning of the square in the Masonic Symbol. In mathematics, the squaring of a circle is calculating a square that would have the same area as a corresponding circle, the quadrature of a circle is simply creating a square that envelops a circle, each side being the same length as the circle's diameter.

Finally, when we divide the circumference (22) by its diameter (7) we get 3.142857, the simple illustration of "pi."

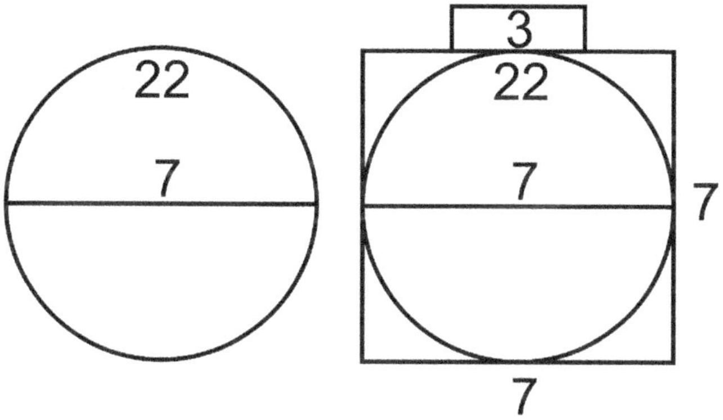

The Square of Sevens
a Mystery Revealed

In the last chapter we explained how a deck of cards in perfect order is quadrated and then laid out in 7 columns and 7 rows with the last 3 cards placed, centered on the top. And as just shown, the Masonic Symbol represents a circle with its diameter being quadrated to form a square with 7 on each of its sides, just like the 7 columns and 7 rows in the Ancient Book of Time. Again, when 22 is divided by 7 we get 3.142857 – the 3 represents the 3 cards placed at the top of the plate and the remainder, 142857 is the Sacred Number.

The Masonic Symbol then, through the symbolism of its compass, square and the letter G beautifully and simply represents the mathematical foundation for the Ancient Book of Time, an old symbolic astronomical system based on the mathematics and symbology behind a regular deck of playing cards, revealing the specific path of each soul, through time and within The God Clock.

More to the Sacred Number

The Sacred Number has been mentioned several times, so what does it really mean and what makes it Sacred? The Sacred Number is 142857, the remainder of the simple version of "pi" (3.142857.) When the deck of cards is quadrated, it is then laid out in the 7 columns and 7 rows, with the last 3 cards in the center on top – and as explained, quadrating the deck 90 times, it forms the 90 plates we progress through in this life – but what is missing? When quadrating the diameter of the circle formed by the compass in the Masonic Symbol, it creates the square that envelops the circle, representing the 7 column and 7 rows in the Ancient Book of Time, but when the circumference of that circle is divided by it's diameter we get 3.142857, And again the Number 3 represents the 3 cards placed across the top and what is left is the remainder 142857 – the missing piece of the puzzle.

When we add the digits of the Sacred Number together (1 + 4 + 2 + 8 +5 + 7), it equals 27 and when we add those 2 digits together (2 + 7) we get 9. 9 is the Number of Mars and the true Number of Man. The Number 142857 then represents action, or life manifest on the material plane, that's why it is Sacred. This explains the missing part within the Ancient Book of Time – Life! What would the 52 symbols and all the 90 quadrations, or plates mean if there were no life? That would be like having an oven that does all these remarkable things, but without any food to put into it to cook, what good would it be?

Another aspect of the Number 142857 is its Symbolic/Card Value – 52 goes into 142857, 2747 times with a remainder of 52 or less. 52 x 2747 = 142844 and when we subtract 142844 from 142857 we get 13, the Solar Value of the King of Hearts. The King of Hearts is the highest card in the suit of love, so it is possible another message of the Sacred Number is that we are here to learn the lesson of love and ultimately, live and evolve to the highest love, finally returning to the source, the center of The God Clock. Finding, or determining the Symbolic/Card Value of a name, or number is simply reducing a number's value either through subtracting 52 from a number until the remainder is 52 or less, or as shown above with large numbers, they are divided by 52 to find the largest number that 52 goes into as a whole number, smaller or equal to the number being calculated. This will be illustrated better later in this chapter.

Back to the Cards and the Joker

In the last Chapter we mentioned the Joker and how to deal with the days of December 31st and February 29th, but we did not get into what the Joker truly symbolizes. The Joker was added to the decks later on (evolving) and, as mentioned before it was (and still in some circles today) believed they were special and could be any card in the deck they wanted. What the Joker truly symbolizes and represents, is illustrated in the Masonic Symbol and that is – the Sacred Number (142857) the simple remainder of "pi", life manifest on the material plane. Which is most truly what the ancients were celebrating on their festival day – they were celebrating the miracle of life itself. When our calendar of 365 days is dived by the Number 7, or Spirit, we get 52.142857. 52 representing both the number of days in each of the 7 planetary periods, along with the 52 Symbols within the deck of cards, and the remainder 142857, the Sacred Number, again the Joker, or life manifest on the material plane.

More to the 7 of Diamonds

The 7 of Diamonds is the 33rd card in the deck, its ruler is Neptune and 7 is also the Number of Spirit. Diamonds represents the element of Earth (as that's from where Diamonds come.) And as shown the capital letter G is given the value of the 7 of Diamonds, as well – Spirit and Earth? Is the Masonic Symbol then trying to tell us that the 7 of Diamonds (the letter G) sitting at the center of the circle drawn by the compass signifies the Soul (Spirit/7) manifesting on the material plane (Earth/Diamonds)? Another interesting note to the 7 of Diamonds is its placement in both the Spirit Plate and the Life Plate. In the Spirit Plate where all the cards are in perfect order from the Ace of Hearts to the King of Spades, it falls in the Saturn Line/Saturn Column – Saturn being the planet that governs Time. In the Life Plate the 7 of Diamonds falls in the Venus Line/Venus Column – Venus being the planet that governs love. Is this revealing to us that Spirit, or the Soulmanifests on the physical plane and through Time to learn, experience and to evolve through love?

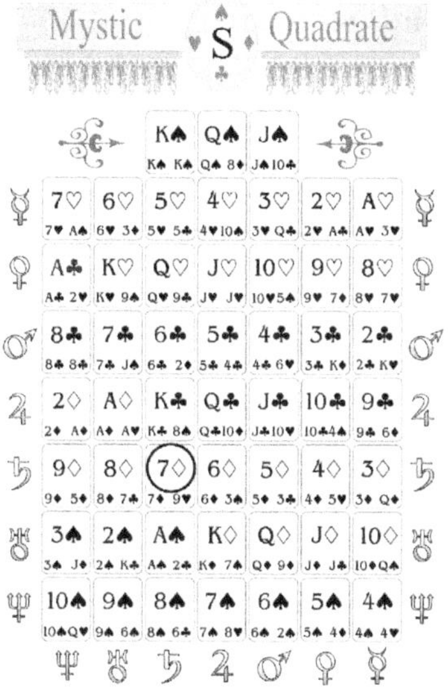

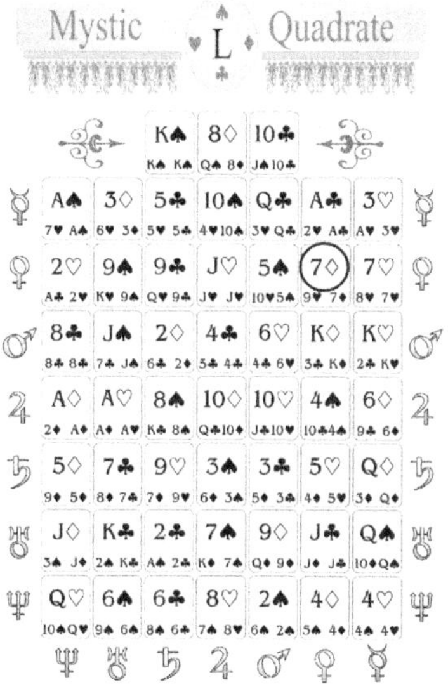

The Alphabet and Numbers
The Ancient Hebrew/Chaldean Alphabet

Before we look at our own 26 letter Alphabet, we must first look into the past to find its probable, mathematical origin and while there are many different graphic illustrations of letters and language, things do at some point, evolve to its more natural path and order. One alphabet that is quite noteworthy is the ancient Hebrew/Chaldean. In the past, wise men looked up and revered the heavens as they knew we were connected to them and that the planets and their motions influenced life here on Earth – they knew "as above, so below." They understood that the vibrations of their letter/words, both in written and spoken form had a meaning and influence upon them. Just as the circle drawn by the compass in the Masonic Symbol has a circumference of 22, representing the Sun, Earth and Moon, the 7 Planets Mercury through Neptune and the 12 signs of the Zodiac (3 + 7 + 12 = 22), it is no coincidence then, that the Hebrew/Chaldean alphabet also has 22 letters.

Our current alphabet has 26 letters, so how does that fit in? Actually our alphabet has 52 letters, 26 small letters and 26 capitals and this was the fundamental basis of discovering the alphabet/card/letter value key. It is quite possible the alphabet divided in two may represent on some level, dark and light, or masculine and feminine – an important duality. So while we think our alphabet just came into being in some random manner, it is more likely that it has a mathematical basis as shown in the revelations within the Masonic Symbol and the Hebrew/Chaldean Alphabet. When the diameter of a circle (the circle being 22, the building blocks of our Solar System) is quadrated to form a square that envelopes the circle with a value of 7 on each side and when the circumference of the circle is divided by its diameter (7), we get 3 plus its remainder and the 3 is placed at the top of the square. When the "Square of Sevens" is squared (7 x 7), we get 49 and add to this the 3 on the top for a total of 52, the actual letters in our alphabet.

Some Remarkable Things

The 22 letters of the Hebrew/Chaldean Alphabet are represented in the Masonic Symbol within the compass and circle drawn by it and through its quadration and division forms not only the structure and form of the Ancient Book of Time, the old symbolic astronomical system based on the regular deck of cards, but is the basis for our alphabet, as well. The discovery then of the basis of the Ancient Book of Time and the alphabet/card/letter value key used to calculate the Card Value of a name and the formation of our 26/52 letter alphabet is not remarkable in and of its self, but that it has been sitting there for possibly thousands of years within the Masonic Symbol, waiting for us to progress and evolve to a point (within an intersection in time) where it can/could be recognized and understood.

There are some other interesting correlations between the Hebrew/Chaldean Alphabet and the Ancient Book of Time.

One is that in the 90 plates, the Cards, for whatever reason are read from right to left just as words written using the Hebrew/Chaldean alphabet are, instead of what is considered normal, from left to right. And, just as in the quadrating of the Cards to form the 90 plates, the missing piece was there was no life manifest in the material plane represented within it. The same with the Hebrew/Chaldean Alphabet as there is no Number 9 within it, 9 representing Man and life manifest on the material plane. Even in the updated version of the Chaldean letter/number correlations with our modern alphabet, there is no Number 9 within it as we, man or mankind are representative of it, we are the life component injected into, or within it.

Now, knowing all of this at this point in time, our alphabet that we use today that we thought was created by chance from the original grunts of cavemen, has actually been written in the stars for thousands of years and we didn't even know it. We were guided, somehow by some unseen natural force within the circles of time, just as we are on our path in life, progressing and evolving to get to this point – in Time.

Names and Numbers

As illustrated. the ancient Hebrew/Chaldean Alphabet of 22 letters is based on the building blocks of our Solar System that influence life here on Earth. Along with that, in our modern alphabet (actually created from the heavens) the letters are given both a card, or symbolic value and a numeric value, each number then is given a corresponding planetary ruler. Interestingly, the Chaldean Number System, just as in the regular deck of playing cards, ends exactly at the Number 52. From this information and knowledge any name, or even a number can be delineated and understood within the Ancient Book of Time, planetarily, or numerically . It should be noted, although not as popular as the Pythagorean Number System, the Chaldean is not only older, but much more accurate – obviously due to its roots in not only the Masonic Symbol, but its relation to the heavens above and the foundation of Time itself.

Using the card/alphabet/letter key, let's take a look at some names and see how remarkable it works. This part of the chapter on symbolic value of names and numbers is not meant as a how to, but to simply illustrate its use and validity. Others works by the author are available that show in more detail how to calculate and interpret these influences within our lives.

The first word we will look at is God.

G = 33 o = 15 d = 4, 33 + 15 + 4 = 52

52 = the King of Spades, highest card in the deck and the card of the Spiritual Master.

How about the name Donald Trump?

D 30, o 15, n 14, a 1, l 12, d 4 30 + 15 + 14 + 1 + 12 + 4 = 76, 76 - 52 = 24
T 46, r 18, u 21, m 13, p 16 46 + 18 + 21 + 13 + 16 = 114, 114 - 52 = 62, 62 - 52 = 10
Donald = 24, Trump = 10
24 + 10 = 34

34 = the 8 of Diamonds, This card sits at the center of the top in what is called the Crown Line in the Life Plate, it is a card of financial strength and those represented by this card either by their birth day or name, like to be the star of the show.

As shown earlier with the Sacred Number, any value over 52, 52 must be subtracted from, or divided by, until the value, or remainder is 52 or less, in order to find a name or number's symbolic value.

How about Cheops, the name given to the Great Pyramid?

C 29, h 8, e 5, o 15, p 16, s 19 29 + 8 + 5 + 15 + 16 + 19 = 92, 92 - 52 = 40

40 is the value of the Ace of Spades, it is known as the card of hidden secrets and mysteries – very fitting.

What about the words "The God Clock"?

T 46, h 8, e 5
46 + 8 + 5 = 59, 59 - 52 = 7
G 33, o 15, d 4
33 + 15 + 4 = 52
C 29, l 12, o 15, c 3, k 11
29 + 12 + 15 + 3 + 11 = 70, 70 - 52 = 18

7 + 52 + 18 = 77, 77 - 52 = 25

25 is the symbolic value of the Queen of Clubs. This is quite interesting, as mentioned before, the Queen of Clubs sits dead center in the Spirit Plate and it is suggested that The God Clock is the center from which all emanates. Queens within The Ancient Book of Time are considered teachers and with the Queen of Clubs at the center, is this another message telling us Time is the teacher? In the Chaldean Number System, which has a different alphabet letter/key, The God Clock adds to a 10/1, 10 representing creation (1, the Number of the Sun/creation) with power (0 the Number of Pluto/power) and 1 representing the beginning, or source.

Mystic S Quadrate

			K♠ K♠ K♠	Q♠ Q♠ 8♦	J♠ J♠ 10♣			
☿	7♡ 7♥ A♠	6♡ 6♥ 3♦	5♡ 5♥ 5♣	4♡ 4♥ 10♠	3♡ 3♥ Q♣	2♡ 2♥ A♣	A♡ A♥ 3♥	☿
♀	A♣ A♣ 2♥	K♡ K♥ 9♠	Q♡ Q♥ 9♣	J♡ J♥ J♥	10♡ 10♥ 5♠	9♡ 9♥ 7♦	8♡ 8♥ 7♥	♀
♂	8♣ 8♣ 8♣	7♣ 7♣ J♠	6♣ 6♣ 2♦	5♣ 5♣ 4♣	4♣ 4♣ 6♥	3♣ 3♣ K♦	2♣ 2♣ K♥	♂
♃	2♢ 2♦ A♦	A♢ A♦ A♥	K♣ K♣ 8♠	Q♣ Q♣ 10♦	J♣ J♣ 10♥	10♣ 10♣ 4♠	9♣ 9♣ 6♦	♃
♄	9♢ 9♦ 5♦	8♢ 8♦ 7♣	7♢ 7♦ 9♥	6♢ 6♦ 3♠	5♢ 5♦ 3♣	4♢ 4♦ 5♥	3♢ 3♦ Q♦	♄
♅	3♠ 3♠ J♦	2♠ 2♠ K♣	A♠ A♠ 2♣	K♢ K♦ 7♠	Q♢ Q♦ 9♦	J♢ J♦ J♣	10♢ 10♦ Q♠	♅
♆	10♠ 10♠ Q♥	9♠ 9♠ 6♠	8♠ 8♠ 6♣	7♠ 7♠ 8♥	6♠ 6♠ 2♠	5♠ 5♠ 4♦	4♠ 4♠ 4♥	♆
	♆	♅	♄	♃	♂	♀	☿	

The White House and the Masons

Let's add in some numbers and take an address such as 1600 Pennsylvania Avenue, the address of the White House, the residence of the President of the United States. Now that we know how the alphabet/card/letter key works in the Ancient Book of Time, we will just show what they are. The Number 1600 calculates out to the Ace of Spades with a value of 40 and the names, or words, Pennsylvania Avenue together calculate out to the Queen of Hearts with a value of 12. Adding 40 to 12 we get a total of 52, the King of Spades, again the highest card in the deck, a very powerful card. Since we have shown the Card System is hidden, or illustrated within the Masonic Symbol and the address of the White House is represented by the highest card in the deck, the card that stands for God, do you think maybe the Masons did actually have something to do with its planning? It is doubtful it was an accident and most likely very well thought out and purposefully designed

More on the Numbers

In Chapter 1 we revealed the unlocking of the Star/Time Code, a mathematical sequence of time based on Saturn, the Earth and the Moon and seasons of those motions and that the Planet Saturn is given the governance of Time. Some believe the Number 7 is the basis of time, but as shown within the Masonic Symbol by the circle and its diameter, 7 represents Spirit, which runs through and divides everything, including time. The Number 7 is powerful and mentioned quite often in spiritual and religious texts.. There are also 7 major notes in the musical scale, 7 main colors in the spectrum of light and 7 chakras within the body. Dr. Brown Landone also mentions within each of the 7 Sacred Temples is a special stone that represents the Seven Powers of God. In the Bible, the Ancient Book of Time is even mentioned as the Little Book of Seven Thunders.

Rev. 10:2 – And he had in his hand a little book open: and he set his right foot upon the seas, and his left foot on the earth. Rev. 10:3 – and cried with a loud voice, as when a lion roareth: and when he had cried, seven thunders uttered their voices.

Neptune being the God of the Seas and associated with the Number 7 and Diamonds being Earth, does Rev. 10-2 and 10-3 illustrate spirit manifest into the physical plane and our journey/path begins in the little book (the Ancient Book of Time) within an astronomical environment as the voices of the 7 thunders would be the planets Mercury through Neptune?

The Number 9 is associated with the planet Mars, the Lord of War, so it is of no surprise it is the Number of Man. If one looks back through man's history, it is quite obvious we, as a race are all about war and conflict. It is the number of disappointment and endings and when using the Ancient Book of Time when Number 7 (Spirit) or Number 9 (Man) cards are derived in a chart from the plates, unless more spiritual in one's approach to life, it can usually indicate challenges and conflict.

While 9 is the Number of endings, 13 is the Number of completion, and as shown earlier, within it is the 4 seasons (1 + 3 = 4) that we must experience in order to fully grown and evolve – Spring/Creation, Summer/Growth, Fall/Harvest and Winter/Death. In the Major Arcana of the Tarot, the 13th Card is the card of Death and Rebirth – the ending of one cycle and the birth of another. The Number 13 is also the basis of all the main time cycles (3 + 6 + 4 = 13) 364 days, 364 years, or even 364 relms, and within 364, are all the other time cycles – 91 (seasonal), 28 (Saturn and the Moon) and 7 (another seasonal.) Again, when these cycles are completed, it is the end of one and the beginning of another – death and rebirth.

Venus is the Planet of Love and 6 is its Number. Have you ever heard the expression "Light equals Love" – let's see what the math says on this one. Light is said to travel at the rate of 186,000 miles per second and when we add all the integers together (1 + 8 + 6 + 0 + 0 + 0 = 15, 1 + 5 = 6) to its smallest number we get 6, 6 being the Number of the planet Venus, the Goddess of Love, so indeed Light does equal Love. Since our purpose here maybe learning and evolving through love and light does in fact equal love, does our journey back to the source literally mean heading back into the light?

The Number 22/52
and the Tarot

Tarot Cards today are not taken too seriously and are more associated with fortune telling than ancient symbolism, but what are they really? Just as there are 22 letters in the Hebrew/Chaldean alphabet and the circle drawn by the compass in the Masonic Symbol will have a circumference of 22, the 22 Cards of the Major Arcana then represent the exact same thing, the 22 building blocks of our Solar System (the Sun, Earth and Moon, the 7 Planets Mercury through Neptune and the 12 Signs of the Zodiac) that not only influence life here on Earth, but are actually the components of creating it. Most Tarot decks today have 78 cards, but we know that the 52 cards, or symbols in the regular deck were created from the quadration of a circle's diameter and that circle being our Solar System. The truth is, there are only 22 Cards in Tarot known as the Major Arcana and their origin is the same as the Hebrew/Chaldean alphabet and Number System. The remaining 56 cards of the Minor Arcana is simply a distorted version of the 52 cards/symbols of the regular deck. The illustration of the Number 22 through the Masonic Symbol and its mathematical expression through quadration and division shows exactly why there are 22 letters in the Hebrew/Chaldean alphabet, 22 cards in the Tarot and why there are exactly 52 cards/symbols in the Ancient Book of Time and why the Chaldean Number System ends at Number 52.

Who Really Were the Masons?

The Masonic Symbol is quite remarkable in itself, showing the mathematical and astronomical foundation of the Ancient Book of Time, the 26/52 letters of our Alphabet, the motion, or mechanics of how they were created, along with life manifest on the material plane and the path, or destiny of each soul while here. There hasn't been a glimmer of this in any Masonic writings/materials that we know of, so where did it come from? It is believed by some that the Masons of today are not the same as the Ancient Masons. The Masons from long ago were understood to know much more mystical, wisdom and knowledge as they were credited as the ones who built some of the great structures such as the Temple of Solomon, one of the 7 Wonders of the Ancient World.

In the 19th and 20th centuries, the Order of the Magi were supposedly the keepers of the secrets namely of the astronomical card/symbolic system, however, looking deeper into the evidence that has been left to us, it is quite obvious that the system known as the Ancient Book of Time, is more likely to originate from the Mayans/Aztecs and/or from the Ancient Masons. And if what Dr. Brown Landone believes is correct, is the true origin of all these wonderful symbols, systems and knowledge from the Mystics in the Eastern Andes, thousands of years ago?

Finally, when we look into a timing device, such as a clock or watch, there are many moving parts such as gears, etc. within the mechanism, which reveals how it works. One has to wonder then, through the illustration of the compass and square, with the letter G in the middle that forms the Masonic Symbol, is it giving us a look into the deeper workings of an incredible mechanism – The God Clock. And, that the letter G represented by the 7 of Diamonds (spirit in earth) placed in the middle is showing us that Spirit and/or the Soul manifests into the physical form at the center of The God Clock, thus beginning our journey on our path, inside the astronomical environment and outward, on our own circles within the Circles of Time.

The End of the Masonic Symbol Chapter....

Chapter Six

The Star/Time Code
The Mayan Time Grids
Where We Are in Time

So far we have seen how and why time unfolds in the manner it does and how this information is encoded in ancient calendars, structures and symbols – and that it all operates within and/or points to The God Clock. In this chapter using the Star/Time Code and the addition of the 7 planetary periods, or divisions of time within them from the Ancient Book of Time, we are going to layout time as we know it, but in larger, proportional blocks, as well. This overlay of time and its planetary influences will be referred to as the Mayan Time Grids. We will take a look at ages/eras, events and places at certain moments, in the past, the present and the future to gain some insight and understanding to where we have been, where we are now and where we will be– in Time. These larger increments aren't necessarily the triggers for events, but create an effect, or influence for them in which to take place.

The Mayan Time Grids

As shown in Chapter 1, time unfolds in a linear/sequential order, in cycles, a proportion manner and all within a circle. In Chapter 2 we revealed the misunderstandings and possible flaws in the current use and perception of the Mayan and Aztec Calendars.

Now we are going to layout time as in Chapter 1, but also include much larger blocks of time and we will call this The Mayan Time Grids.

Using a year as an example, there are basically 4 main cycles of time, the 364 day cycle with its 4 seasonal cycles of 91 days and a 28 day Lunar cycle with its 4 seasonal cycles of 7 days. Each of these 4 cycles will have 7 planetary periods, or divisions time within them. Each of the 7 planetary periods will be given a planet that governs that period; the first planetary period in any cycle is always Mercury, the second is Venus, the third is Mars, the fourth is Jupiter, the fifth is Saturn, the sixth is Uranus and the seventh Neptune. The same order, or sequence as reflected above in our Solar System.

The 364 day cycle will have 7 planetary periods of 52 days (7 x 52 = 364)

Each of the 4, 91 day cycles will have 7 planetary periods of 13 days (7 x 13 = 91)

The 28 day cycle will have 7 planetary periods of 4 days (7 x 4 = 28)

And each of the 4, 7 day cycles will have 7 planetary periods of 1 day (7 x 1 = 7)

Taking this to the next larger increment from days to years we will have :

The 364 year cycle will have 7 planetary periods of 52 years (7 x 52 = 364)

Each of the 4, 91 year cycles will have 7 planetary periods of 13 years (7 x 13 = 91)

The 28 year cycle will have 7 planetary periods of 4 years (7 x 4 = 28)

And each of the 4, 7 year cycles will have 7 planetary periods of 1 year (7 x 1 = 7)

Taking this to the next larger increment from years to relms we will have:

The 364 relm cycle will have 7 planetary periods of 52 relms (7 x 52 = 364)

Each of the 4, 91 relm cycles will have 7 planetary periods of 13 relms (7 x 13 = 91)

The 28 relm cycle will have 7 planetary periods of 4 relms (7 x 4 = 28)

And each of the 4, 7 relm cycles will have 7 planetary periods of 1 relm (7 x 1 = 7)

As shown in previous chapters and as just illustrated in the breakdown of the 3 different increments (days, years and relms), this is how time unfolds and always has. The smaller cycles happen precisely within the larger ones and are perfectly proportional, as well.

364 relms (132,496 years)   364 years (132,496 days)
364 days
91 relms (33,124 years)   91 years (33,124 days)
91 days
28 relms (10,192 years)   28 years (10,192 days)
28 days
7 relms (2,548 years)   7 years (2,548 days)
7 days

There are even larger blocks of time that would be perfectly proportional to what we have laid out, but might be so big that it would be hard for us with our perception of time to understand, or see its influence. The relm cycles are large enough and we will use that to start with and work down from there. We are going to look at a few significant events that have transpired, as well as, what we will call ages and the influences present at that time. Using the Mayan Time Grids, we will be able to see what different cycles, seasonal influences (Spring/creation, Summer/growth, Fall/harvest, Winter/death/completion) and specific planetary periods that were in effect at the time of the event, or age.

Here are the 4, 7 Relm Cycles within a 28 year Relm Cycle that we are currently in. We are using the start of the Mayan Calendar as a point origin. From these increments we will be able to delineate time into even smaller influences.

As an example in the Mayan Time Grid table we can see we started a 7 Relm Seasonal Fall Cycle in the current 28 Relm Cycle in 1982 and we also started a 364 year Mercury period in 1982.

The Four 7 Relm/Seasonal 2548 Year Cycles within the 28 Relm/10192 Year Cycle							
And the Seven 1 Relm/364 Year Planetary Periods within the 7 Relm/ Seasonal 2548 Year Cycles							
Spring ♥ 3114 BC		Summer ♣ 566 BC		Fall ♦ 1982 AD		Winter ♠ 4530 AD	
Mer	3114 BC	Mer	566 BC	Mer	1982 AD	Mer	4530 AD
Venus	2750 BC	Venus	202 BC	Venus	2346 AD	Venus	4894 AD
Mars	2386 BC	Mars	162 AD	Mars	2710 AD	Mars	5258 AD
Jupiter	2022 BC	Jupiter	526 AD	Jupiter	3074 AD	Jupiter	5622 AD
Saturn	1658 BC	Saturn	890 AD	Saturn	3438 AD	Saturn	5986 AD
Uranus	1294 BC	Uranus	1254 AD	Uranus	3802 AD	Uranus	6350 AD
Neptune	950 BC	Neptune	1618 AD	Neptune	4166 AD	Neptune	6714 AD

Now let's take a look at some moments in time to actually see the Mayan Time Grids in action and what they have to say. Showing all the seasonal and planetary periods in the 3 events that are going to be explained may seem a little overwhelming, or complicated, but just be patient, as we are just trying to show the story being told through the influences present, regarding the event – within time. It is a story worth telling. The focus of this book is more on understanding the mechanics of time and how we relate and experience it within The God Clock, the full layouts and use of the Mayan Time Grids will be in a future work.

The Birth of Jesus

Jesus, the Son of God, who it is said died for our sins on the Cross was born between 1BC and 1AD which is recognized as the starting point of the Gregorian Calendar (the one we use today), even though this calendar was not implemented or commissioned until 1582 by Pope Gregory XIII. And it is why many times our year is literally referred to as "The Year of Our Lord." There are other theories as to which year Jesus was actually born based on both astronomical and historical evidence, but for now we will stick with the 1BC/1AD year and it may actually show it is the correct year after all. For ease of reference we will refer to 1BC/1AD time marker as the Zero (0) point.

Looking at our Mayan Time Grid table the 0 point falls in the 2nd Summer 7 Relm Seasonal Cycle (2,548 years) of a 28 Relm Cycle (10,192 years). Falling in a Summer Season, this 7 Relm or 2,548 year period would represent a time of growth in which we were in until 1982 when a new (Fall) 7 Relm/2,548 year Season started. The 364 year planetary period active for the 0 point was Venus from 202 BC through 162 AD. Venus is the Planet of Love which is what Jesus was here to teach us about. 364 years is a cycle of its own and would have 7 planetary periods of 52 years. Starting at 202 BC and adding to that in 52 year increments, the 4th one would start in 46 BC and would be governed by the Planet Jupiter (202 BC Mercury, 150 BC Venus, 98 BC Mars, 46 BC Jupiter (0 point) and 6 AD Saturn, etc.) Jupiter is known as a benevolent, or good influence and has an expanding nature and would do so over this 52 year time period – another fortunate time for Jesus to be born and his apparent influence upon us.

During this Venus Period of 1 Relm, or 364 years starting in 202 BC, there would be 4 Seasonal cycles of 91 years and 7 planetary periods of 13 years within it, as well. Progressing up in 91 year increments from 202 BC, we would start the 3rd or Fall Season in 20 BC. (Spring 202 BC, Summer 111 BC, Fall 20 BC (0 point) and Winter 71 AD.) 20 BC then would be the start of the 13 year Mercury Period and the 13 year Venus Period would start in 7 BC – the 0 point would fall within this time frame. Being in a 91 year Fall Season it would be a time of harvest, building on something that was created and grown from the previous 2 Seasons (Spring 202 BC and Summer 111 BC) but on a larger scale still in the growing stage as the world was in the Summer Season of 7 Relms, or 2548 years. Falling in the Venus Period again points to love, the main lesson of Jesus.

Also, within the 1 Relm or 364 year cycle, there will be 13, 28 year Saturn Cycles. Counting up from 202 BC in 28 year increments, we would start the 8th one in 6 BC. The 7 planetary periods within this cycle are 4 years in length and the Mercury Period would start on 6 BC, adding 4 years to this we get to the Venus Period which would start in 2 BC and the 0 point would fall within this time period – Venus the Planet of Love again.

Finally, we get into the 4, 7 year Seasonal cycles within the 8th 28 year Saturn cycle and these will have 7 planetary periods of 1 year. With the Mercury Period starting in 6 BC, we would just go forward a year at a time. Mercury 6 BC, Venus 5 BC, Mars 4 BC, Jupiter 3 BC, Saturn 2 BC, Uranus 1 BC, (0 Point here) Neptune 1 AD. The 0 point then would be the start of the Yearly Neptune Period and 2 AD would be the start of the next 7 year Seasonal Cycle. Neptune is the Planet associated with the Number 7, the Number of spirit and governs spirituality. Neptune/7 also is connected with sacrifice of which Jesus did. As the story goes he made the ultimate sacrifice by dying on the cross, for our sins. It is believed Jesus was born during the Age of Pisces, which is associated with the Planet Neptune. This could be another environmental influence magnifying the effects of being born within a Neptune Year.

Looking at all the influences at play at the time of the 0 point using the Mayan Time Grids, one could make the case that through Jesus, we were to grow (large Summer Season) and build upon (smaller Fall Season) from his message of love (3 Venus Periods) to expand and receive good (52 Year Jupiter Period) through his spiritual influence and teachings, along with his sacrifice (Neptune Year.) Now looking at time in this manner, it may very well be the 0 point is the actual birth year of Jesus.

The Return of Jesus?

If and when Jesus does return, we have to consider a few things. If He comes back during the same influences as the last time within the next 28 Relm, or 10192 year cycle it wouldn't be until the year 10192. Quite a ways away, but interesting it is the same number of years as in a 28 Relm Cycle – 10192. Or would he come back in the next 7 Relm Seasonal Cycle of 2548 years, which we are now in and started in 1982? Again, are the influences the same, or similar, or will/does He come back under different Seasons and Planetary influences to teach us something different each time, like a progressive, evolutionary, spiritual advancement? We will see when the Time comes.

Just a Story

Some believe that the story of Jesus, is just that – a story – and that He was never here at all. Even if that were actually true, the 0 point time marker is still quite substantial and had to be chosen for some very important purpose, within time.

Note: In the birth of Jesus a more detailed illustration of how the time increments were determined and used in the Mayan Time Grids, to show how they worked. They may at first, seem complex in explaining them, but they are actually quite easy to use. From this point on to save time and to keep from boring you, we will just simply state what the different Cycles, Seasons and Planetary Periods are in play at the time of the event. A detailed set of Mayan Time Grid Tables will be available later by the author.

The Great Pyramid

There are many theories on when the Great Pyramid was built, including one that it may have been built in, or around 10,500 BC. This theory is based upon the upon the picture of the heavens above at that time, in the Age of Leo and that the layout on the Giza Plateau of the 3 pyramids and the Sphinx reflected that on the ground – "as above, so below." Even if the Great Pyramid was not built during that time, it may point to something important regarding that time, alignment, or to a connection to somewhere out in the galaxy. And if it really was built in the distant past, what about the possibility it may have been built in the year 10,192 BC, the number of years in a 28 Relm Cycle?

According to Dr. Brown Landone, the Great Pyramid was built in 2881 BC and we will go with that for now, as his chronological time line on the 7 Sacred Temples and their details here on Earth, appears to be based on something perhaps valid. 2881 BC in the Mayan Time Grids falls into these Seasonal and Planetary periods; the 1st (Spring) 7 Relm (2,548 years) Season, a 1 Relm, 364 year Mercury Period, a Saturn 52 Year Period, a 13 year Saturn Period within the 3rd (Fall) 91 Year Season, a 4 year Mars period within the 9th 28 year Saturn Cycle and a Mars Year within the 2nd (Summer) 7 year Season. Through the 3 Seasons in play we can ascertain that it was created (Spring) to harvest, or build on (Fall) in which to learn and grow (Summer). The first planetary period of that 364 year Cycle is Mercury, representing a time of information, communication and knowledge – the Message of the structure. The 52 year and 13 year Planetary Periods were both Saturn, Saturn is the teacher, but also is the planet given the governance of Time and maybe in this case laying out for us the structure of Time. The 4 year and Yearly Periods were both Mars. 9 is the Number of Mars, the Number of Man and what the Sacred Number (142857) reduces to when all the integers are added together until the remainder is a single Number. Through the Mars influences, the builders may be telling us this represents man himself. Also, Mars is the Planet of energy and action and it was within these smaller increments for timing purposes it was to be built in order to get it going and complete it.

A Hebrew Connection?

Another aspect is that 2881 (the year the Great Pyramid was built) adds to a 19/1 (2 + 8 + 8 + 1 = 19, 1 + 9 = 10, 1 + 0 = 1.) In the ancient Hebrew Calendar the Sun and Moon were used to track time, and in the documentary "The Mysteries of Chaco Canyon" the native peoples went to great lengths in and on their structures to precisely track the movements of the Sun and Moon. It takes 19 years for the Sun and Moon to complete a cycle, falling on the same month and day as it did 19 years before. Is this another part of the message they left for us by the year in which it was started –

the tracking of time? Using the Card Value technique the Number 2881 calculates out to the 8 of Clubs, 8 the Number of Saturn and Clubs representing the Element of Air – Saturn in Air. Saturn, the planet that governs time and the Element of Air representing things that are light and quick, information, knowledge, etc. Again, if it was built in the year 2881 BC and through its numeric and symbolic value, was it done intentionally to teach us about time? Another interesting point is that it was built in the 1st 364 year Mercury Period after the start of the Mayan Calendar. Things happen for not only a reason, or purpose, but at that specific time, as well – maybe we were starting an Era, or an Age where it was time to learn and know (Mercury), or at the very least, get the information out there for us to connect to within another circle of time. The Great Pyramid could have been constructed in another year, before or after 2881 BC, but a case could be made from the information given, 2881 BC was when it was built, or started to be built.

9/11/2001

Now let's take a look at something more recent – the event that occurred on September 11th 2001, the attack on the Twin Towers in New York, City. Just like the Great Pyramid, we won't get into the details and just state the influences present at the time of the event within the Mayan Time Grids. They are a 3rd (Fall) 7 Relm (2,548 years) Season, a 1 Relm, 364 year Mercury Period, a 52 year Mercury Period, a 13 year Venus Period within the 1st (Spring) 91year Cycle, a 4 year Uranus Period within the 1st 28 year Saturn Cycle and a Neptune Year during the 3rd (Fall) 7 year Season. The first large Seasonal increment is 7 Relms, or 2,548 years and is the 3rd or Fall Season, during this time we are, on a large scale harvesting/building things from what we have created and grown from before, taking it to the next level so to speak. Next is a 91 year Spring Season, this indicates on a slightly smaller scale will be creation or desire for new things and finally the 7 year Season is also a Fall one, from these creations within a larger cycle, we will be harvesting or building from that on a smaller scale.

The first 2 planetary periods in play at the time, the 1 Relm/364 Year and 52 year one were under the influence of Mercury placing the focus on communication, information, as well as things happening quickly. The next is the 13 year period which is Venus – besides love, Venus rules money, which many believe is what this event was about (and many other things within this time frame.) Going smaller we have a 4 year Uranus Period, Uranus represents the unexpected and it seemed very much so. And it happened in a Neptune Year, Neptune ruling illusion and/or something hidden – a secret – with many believing, and possibly rightly so, that something was being hidden. Neptune also rules oil another aspect of this event.

Finally the year 2001 adds to a 3 (2 + 0 + 0 + 1 = 3) the Number of Jupiter which rules expansion and magnification – something for good or bad was definitely magnified. Jupiter also governs wealth and business, two things greatly affected on that day. Symbolic wise, 2001 calculates out to a Queen of

Clubs, which in the Life Plate falls in the Mercury Line/Mars Column – sudden or quick action/conflict. Since it is a Jupiter year (3) the secondary card here is the Jupiter Card of the Queen of Clubs, which is the Ace of Spades, the card representing powerful transformations and the Symbol of Death. By it's placement in the Life Plate in the Mercury Line/Neptune Column it is also considered the Card of hidden information, mysteries and secrets, which may have been enhanced, due to it being a Neptune Year. Sometimes things are not quite so clear and we have to put the pieces of the puzzle together to see them as they really are.

The Ages

We have looked at certain events and structures and the cases to be made regarding them within the Mayan Time Grids from large blocks of time down to the Year, next lets look at some larger influences on their own and see their effects upon us in time. More specifically, we will look at a few 52 year Planetary Periods as that is an increment we should be able to see its influence and effects during that time. These were mentioned in previous chapters, but now that we have seen the Mayan Time Grids, let's take a closer look at them.

The Age of Uranus
The Age of Industry

In the year 1878, we started a 52 year Uranus period and it ran until 1930. Uranus is the 7th planet from the Sun (including Earth) was discovered in 1871 and orbits the Sun in approximately 84 years. It governs areas such as industry, labor, technology and electricity. It also represents things unusual, or unexpected possibly due to the fact that it rotates in the opposite direction of most other planets..

It is during this 52 year period is when we became industrialized and producing much more on much larger scale. This created a need for an expansion and a demand for industrialized labor. Along with this came the start of Labor movements, in which some people even lost their lives fighting for workers rights and better working conditions. Technology was on the rise during this time and many machines and devices were invented, as well as, advancements with electricity. This was also a time for great, unusual minds like Tesla and Einstein with their remarkable inventions and theories, including Einstein's Theory of Relativity – $E=mc^2$. A momentous age and quite fitting for the Planet Uranus.

The Age of Neptune
The Age of Drugs, Oil and Dreams
The Born Again Movement And the End of an Era

After the 52 year Uranus Period in 1930, we started a 52 year Neptune Period and this ran until 1982. The Planet Neptune is the 8th planet from the Sun, its discovery was in 1846 and has an orbit around the sun of approximately 165 years. Neptune is associated with such areas as drugs, oil and dreams. During this time there was a great expansion in the discovery and use of vaccines, which we still use today for the betterment of people's health. Although drug use has been around for centuries, it really came in vouge in the 1960's and 1970's during the creation of the Hippie culture and shortly after, sex drugs and rock and roll. Even though oil was not discovered, or its first use was not during this time frame, a case could be made that it was during this period that we developed a much larger use and dependence for it and as some think, became addicted to it. It will probably take the right intersections, planetary influences, choices and will to finally get off of it and break the addiction. Neptune also rules dreams and it is under this influence and time, the American Dream was formed. Of course in 1982 we ended the 52 year Neptune Period and looking back it is probably when the American Dream actually ended.

James Truslow Adams in 1931 (the Age of Neptune started in 1930) on the American Dream "life should be better and richer and fuller for everyone, with opportunity for each according to ability or achievement" regardless of social class or circumstances of birth. And during this period it was.

Since Jesus was supposedly born during the Age of Pisces the Zodiac Sign associated with the Planet Neptune, it is no coincidence that the Born Again Movement's "Golden Era" happened in the 1970's – falling in this 52 year Neptune Age and within the 13 year Neptune Period of the 91 year Season in effect during that time, as well.

In 1982 we completed both a 7 Relm 2nd (Summer) Season of 2,548 years that started in 566 BC and a 364 year Neptune Cycle that started in 1618 AD, this then would be the end of a large (7 Relms) and not quite so large (364 year) Era. 2,548 years of growth and learning and capping it off with 364 years of Neptune, the planet of dreams, spirituality, sacrifice and unfortunately escapism and drugs.

The Age of Mercury
The Information Age
The Start of a New Era

The Mayan Calendar was thought to end on December 21st in the year 2012, however, nothing has seemed to change at that point. We did however experience an end to an Era in 1982 as just mentioned. From the start of the Mayan Calendar in 3114 BC, while not a larger block of time like a 26,000 year increment (what ever that would represent), we are about half way through a 28 Relm Cycle which lasts 10,192 years.

In 1982 we did start a new Fall 7 Relm Season of 2,548 years and a new 364 year Mercury Period and within that a 52 year Mercury Period, which is where we are now.

The Planet Mercury is the closest planet to the Sun and has an orbit of approximately 88 days. Its area of governance is information, communication, knowledge and transportation along with younger people and pets. During this time we have seen children and younger people's issues come to the forefront, both for good and bad – in some ways they are better protected and in others over protected. Regarding small pets the formation of shelters and other means to help them have greatly increased during this time. We have also seen big changes in cars, as they have become much more than just a vehicle to go to the store or work and our communication devices are linked, or synced right into them. But the big deal with this Mercury Period so far is in information and communication, we have been buried in all kinds of information with cable and satellite tv, 24/7 cable news, the Internet, cell phones, the importance placed on education, online college courses, YouTube, etc. With an obvious explosion of all things Mercury, one has to wonder since we are also in a 364 year Mercury period within a larger increment of time (a 7 Relm/2548 year Season) is that influence magnified in some way?

While we are just showing the larger influences at play here, if one wanted to, they could dig deeper into the Mayan Time Grids to find a possible Uranus/Technology presence as much of what is being built today is technology based; however, the overall theme is communication, information, travel/vehicles and Mercury being the conduit in which it is coming through.

Since we have entered a new Age/Era, it could be why so many feel a sense of uncertainty. After all, we are in territory rarely experienced by people here on Earth, at least to us at this point in Time – going from one 7 Relm/2548 year cycle into another and starting the 2nd half of a 28 Relm/10,192 year Cycle. For those of us old enough still living in the residue and the "used to it" feeling of the last Age/Era (pre 1982), it may not have set in right away as we entered the new one, but now on a deeper level we know we are on a new course through time and we are not sure yet where it is taking us. For those younger and born within it, they may not have this same sense.

The Age of Venus
The Age of the Woman
Into the Future

Looking a little into the future we will enter a 52 year Venus Period in the Year 2034. The Planet Venus is the 2nd planet from the Sun, it is the 2nd brightest object in the night sky and has an orbit around the Sun of about 225 days. Venus rules beauty, music, the arts, money, food, indulgence and naturally the woman. During this time we might see a change in money in some way, maybe a unified system for all nations, for good or bad, or possibly a lessening of disparity between the haves and the have nots.

True beauty, art and music may once again take the forefront, instead of the pop/manufactured stuff we get today. And indulgence – for the worse of course.

Food is very important and another area ruled by Venus. Today much of our food is processed, comes in a box or can and it is not good for us in many ways. Along with this much of our food is becoming genetically modified, another negative regarding the food supply and its effects upon us. The Age of Venus should also bring our food to the forefront, possibly at times shortages of it (within a smaller Saturn period?), but possible changes will be made to improve it, to be more nutritious and natural and in turn we will be much healthier for it. The Medical/Pharmaceutical establishment wouldn't be very happy about this as it is more profitable with people being sick, but that is for another Planetary Period to deal with, or within this one.

And finally women – will it be during this time that they take their rightful place as equals along side men? Will they finally receive equal pay for the same work? Will the barbaric abuses of women in foreign lands finally come to an end? This is interesting because as we get nearer to a new age, we start to perceive and/or feel it on some level and this could in a partial way explain the current war on women – the powers that be know and feel change is on the horizon and they do not want it to happen, therefore they are not only doing their best to keep it from happening, but trying to turn back the clock, to make even harder for it to occur at all.

Even though we aren't quite there yet, we can already see events and experiences starting to happen to set up what needs to be in order for the changes and events to take place in the Age of the Woman. Very soon we may see the 1st woman President of the United States and that doesn't imply that it would be good or bad, just that it would happen. Even if a woman at this point in time doesn't get the nomination from their party, or elected President, it is setting it up for things to come. Some of the changes for women might not be easy during this time, but most likely things will be different for Women in general after this 52 year Period.

Finally, Venus is the Planet of Libra which also rules through balance and peace. This may be a time where peace finally reigns over the world for many years – lets hope, if not sooner. We actually enter a larger 364 year Venus period in the year 2346 and another 52 year Venus period within that, in the year 2398. Might that also be a time of great and lasting peace on the planet Earth?

Mars
The Lord of War

It should be noted that within the new 364 year Cycle started in 1982, we would also experience 4 - 91 year Seasonal Cycles and each of those having 7 Planetary Periods within them. The first 91 year Season that started in 1982 would be a Spring or Creation Season. From 1982 through 1994 would be a 13 year Mercury Period, from 1995 through 2007 would be a 13 year Venus Period and from 2008 through 2020 would be a 13 year Mars Period. At the time of this writing, we are in the 13 year Mars Period within a Spring/Creation Season. Looking at the world at this point, one could easily see the powers that be are very busy creating a lot of wars/conflicts at this time.

Was the Destiny of America
written Thousands of Years Ago?

Over the centuries, great countries, regimes and empires have risen and fallen and in more recent times, the United States, a new country declaring independence on July 4th, 1776 has risen to great heights. The Symbol that represents this day is the Jack of Diamonds, initiation into higher worth and values. In the Life Plate if falls within the Neptune Column/Uranus Line – the Spirit of the Humanities. While this sounds wonderful and will be an intersection within time to hope for as we head into the future, the truth is America in the larger scheme of things is a very young country. Jacks can represent a youthful and immature persona and it takes time to mature and grow from lessons learned along the way. America is surely not acting in its highest good during these times (although it always claims to be), but the potential is there for it to truly shine.

Regarding its destiny, there are some that believe when America attains the number of 52 states, its true destiny will be revealed. Interesting as 52 is the number of working symbols within the Ancient Book of Time and the

Hebrew/Chaldean Numeric system ends at Number 52. Not to mention its incorporation mathematically into the Great Pyramid, the Mayan Pyramid and the Masonic Symbol – pointing to another Masonic origin of America.

Another possibility for the destiny of America is the origin, or starting year of the Mayan Calendar, which was 3114 BC. Using the technique to determine the symbolic value of a number, the year 3114 calculates out to the Jack of Diamonds, which is the card that represents the United States. Is it then possible a specific point in time was chosen, not only to start an essential calendar, but for other important events and ideals to take shape and form at particular intersections within circles of time in the future – and from these it would create the necessary vehicle in which to evolve and grow. Was America's destiny written in time thousands of years ago? When the day comes it reaches 52 states will significant intersections within time occur and its true destiny be revealed? We shall see.

Note: America is not the only country on Earth and within circles it is connected to all other countries and governments in the world, for both good and bad. It just has the potential at this time to be the light of the world and humanity. Hopefully as we reach those important intersections in time, the right choices will be made, or actions taken regarding the good of all.

Here We Are

Using the Mayan Time Grids which actually evolved from the Star/Time Code in a simple, yet proportional way, it gives a different look into larger increments of time and their effects at various times in our history and what is possibly to come. From great cycles/circles of time we may not really see or notice, to smaller ones that are very obvious as to what they are about and their influence upon us. And just as our now seemingly small and personal circles of time unfold within our lives along our path, we are also along for the ride within these larger ones showing another aspect of The God Clock, and how precisely, magnificently and proportionately time unfolds in its many ways within it.

The End of Another Time Chapter
Goodbye 1982

Chapter Seven

The God Clock
The Final Chapter

Time Within the God Clock

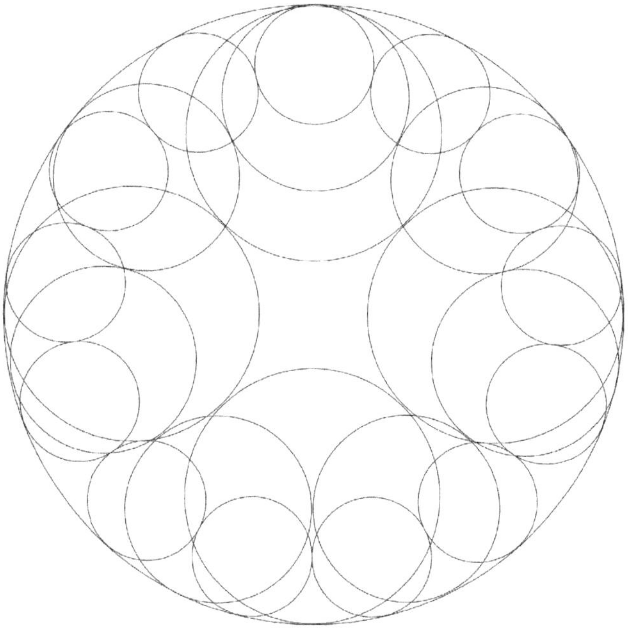

The graphic on the beginning page of this chapter, is the same graphic that is on the front cover of this book and represents the circles of time that we experience. Just as in a fine watch, or clock with precise gears that travel on their course inside the mechanism, setting the hands on the face in motion to mark time as we go through it, like mile markers going by as we are driving down the highway – these circles of time are the gears or wheels moving so effortlessly, precisely and proportionally within the mechanism known as The God Clock.

The large outer circle is the completion increment of 364, the next size smaller are the 4 Seasonal increments of 91, going down a size we have the 7 Planetary Periods of 52 and finally the 13 small circles representing the 28 unit Saturn/Lunar motions. While downsized (proportionally, of course) to fit into this book, the original graphic was created in a computer drawing program, the first big circle was drawn with a 7" diameter having a circumference of 22." The other circles were drawn and manipulated, until they would fit inside the large circle according to their number (4, 7, or 13) and just touching each other. These just happen to be the 1st four numbers of the Telios Proportions as explained in Chapter 1 (1, 4, 7 and 13.) This eloquently illustrates that the numbers on which time is based and unfolding, are not only proportional in both linear and cyclic environments, but that it might even be proportional in the very essence of its fabric.

An Art Gallery at the Intersection of 7th Avenue and 22nd Street?

Looking at this graphic we see a beautiful blend of proportional and symmetrical circles, lines and shapes, just as in classic art that has stood the test of time (is time supposed to be graceful and is there a test, as well?) Another important point within these circles is the intersections between them. These would represent points in time where different circles of time come together and could signify, be the trigger for, or be important markers of some kind. If we look at time itself unfolding within and through these circles and using the 0 point as suggested in the last chapter for going back and forward in measure, would their intersections point to important dates/times, events and experiences man and the Earth are supposed to undergo, to learn and from which to grow?

And not only do we have time itself unfolding within these circles, but we have our own unfolding in the same manner that started the day we were born. Each circle or wheel in time, including ours would have gears that mesh together, so the day on which we are born we are integrated into time in whatever point it is at and at that moment we are now part of the mechanism and our journey begins in this life within The God Clock. It would be like time and the people already in it are on the interstate moving along, and when we are born, we come down the entrance ramp, blending in right with the flow. So how would these circles actually mesh together, what would these gears be made of, or based on? They are precisely made up of days, weeks, months, years, relms, etc. – measurements of Time. The gears of one circle, or wheel are made simply from the increments of a smaller circle, or wheel – perfectly and proportionately interlaced, incredibly efficient and in essence, creating the fabric of time. If time is a fabric, would the intersections within it actually be what is known as – a stitch in time?

Note: Dr. Brown Landone speaks of the importance of the intersections of circles drawn to the Telios Proportions within each of the 7 Sacred Temples around the world and their importance regarding specific dates and man's journey, however, what he is talking about is done by drawing these circles within the construct of a square or rectangle. What is being proposed in this work is that instead of the circles being drawn inside certain parameters, or within defined edges, we are using the center of The God Clock for the point of origin, and drawing the circles from that and then expanding, or unfolding outward from there (in proportion) using the 0 point (1 BC/1 AD) as a time marker, a place from where we can start, or measure. The intersections then, being formed naturally within time, as illustrated by the graphic at the beginning of this chapter. This doesn't mean that the circles drawn to Telios Proportions within the 7 Sacred Temples are wrong, it would simply suggest, there is more to the story.

A Ripple in Time

As demonstrated in the event between Michael Brown and Darren Wilson in Ferguson, MO, not only are there intersections between time itself, but between a person and time, also between two or more people, and along with that, happening within the circles of/or at a geographical position. The possibility then exists that significant events will and do transpire when specific circles of time do in fact intersect and the number of people involved with that event, either directly or indirectly, is based on the degree of closeness of their circles intersecting to that point/event. Depending on the number and size of these intersections and the vastness and degree of connections to them, might this be what we think of as – a ripple in time?

Connections Through Time
Rinse Cycle Repeat

Truths known hundreds or, thousands of years ago were disregarded or destroyed for what ever reasons (usually bad ones), then at a later time re-discovered. This knowledge and understanding was set out in a circle only to return at a later point in time, possibly at an intersection within an Era or Age in order to create the environment for it to manifest and be understood again. This is a powerful truth in itself – even though the past maybe long forgotten, or truths and/or knowledge seemingly destroyed, we are still connected to it through the circles of time.

And just think – right now – at this moment in time – someday – we will be the past, the future will be connected to – at that moment in time.

Learning these connections and intersections to know what is to come and what to expect on our journey is important, as cycles, seasons, etc. do repeat themselves. While theses cycles can reoccur in large time increments, let's take a look at a much smaller and simpler one to illustrate the point. Within the Ancient Book of Time, as described, the deck of cards is quadrated 90 times, then falls back into perfect order. Along with that a couple of interesting anomalies are happening within the mathematical, sequential layouts created from the quadrations. First, two pairs of cards simply switch positions in each successive plate and secondly, the 1st 45 quadrations are the same as the 2nd 45 (remember there are 90 total) and the only difference is that these two pairs of cards are in opposite positions. (The two pairs of cards are, the 2 of Hearts/Ace of Clubs and the 9 of Hearts/7 of Diamonds.) With this in mind, lets get back to the point.

On March 13th 2013, Cardinal Bergoglio was elected to and became Pope Francis at the age of 76. Since the first half of the quadrations are the same as the last half (except for the two pairs of cards that have switched places) Plate, or age 76 would be the same as Plate or age 31. One has to wonder then, what happened in Pope Francis' life during this same time when he was 31 years old? Bergoglio worked as a chemical technician and as a night club bouncer – then – almost to the day entered the Society of Jesus to begin his studies to become a Jesuit Priest. This illustrates what happens in the past at some point will happen in the present and happen again in the future, as well.

Ecclesiastes: 15 – Whatever is has already been, and
what will be has been before

The Native Americans Have Been Right All Along

One of the Spiritual truths held by Native Americans is that everything is a circle. But now, knowing and understanding how time unfolds and actually seeing it in a graphic format, we know and see it is true. However, it is telling us something else that is very important that we have forgotten – we are all connected and through recognizing this simple truth, we should be working together instead of in the greedy, selfish manner in which most exist in today. Another circle that we have deep connection to is to the Earth, as it is one of the vehicles in which we are traveling on/in during our journey within the circles of time. We need to take better care of her, as we cannot just trade her in for a new one like a car after we have abused and neglected it.

Another powerful aspect within Native American traditions, is their honoring of the Four Directions. This is quite interesting as in the Astronomical Card System (the Ancient Book of Time) the deck is Quadrated, or thrown to the Four Directions and this quadration, or structure of time is also mathematically hidden within the Masonic Symbol. In signifying the Four Directions are they really on some level honoring the formation and/or the creation of and our journey through time within the physical plane?

Earlier in the book we have shown the importance of the relationship between a circle (22) and its diameter (7) and how it beautifully illustrates that Spirit (7, the diameter of a circle) runs through and divides everything. The Native Americans have already known this for a long time, revealed by their belief and honoring the spirit in everything, including the heavens above. The Europeans that came to this country thinking their beliefs were truth and that the Native American beliefs were primitive and foolish. Now over 500 years later, we are just catching up to the wisdom and knowledge they have held sacred for thousands of years. Funny – we feel we are so much more cultured and educated than they were (or are), yet our higher math and science, the very things we think make us better, will prove that they have been right all along.

Pretend you are a Native American and take another look at the circles of time graphic at the beginning of this chapter – doesn't that almost look like a Dream Catcher? Could they have really understood the circles of time and the intersections within and is this also what they were symbolizing in this woven circle? Were, or are they trying to catch beneficial intersections within time, attempting to connect with them? After all, that's what they were built for – catching dreams.

The Religion of the Clock
Welcome to ClockMart

We can be sure when science one day proves the existence of The God Clock (at some important intersection in time), some religion will step up and contend to know all the answers about time and help guide us through it. The God Clock is non denominational, no religion owns it and while some religions may want to claim it, define it and try to set our boundaries within it, the Native Americans will have it right again, by simply honoring it.

The next step of course is for a case to be in front of the Supreme Court where by a corporation's filing of a patent to own The God Clock will be decided. And when the corporation wins the case, you will then have to go to ClockMart and buy your time. Upon entering the time store, the greeter will look at his or her watch, then look up and say "Welcome to ClockMart." No one owns, or will ever own time itself and you won't even be able to buy it at the fabric store.

The Importance of Now

We have often heard by spiritual experts and motivational speakers of "living in the moment", or "existing in the now," however, knowing that time unfolds in circles that is not only happening right now, but is both connected to the past and connected to the future.

This gives us quite a different perspective of what now really is. Past, Present and Future are basically one in the same.

What and where we are at this moment, where the universe and our Earth are at this moment is directly connected to not only what happened 1 second ago, but also to what happened thousands and even millions of years ago. To carry that further in the other direction, where and what we are doing right now at this moment, is directly influencing and connected to where and what we will be, in the future.

Who we are at this moment is based our what has happened to us before, including not only our childhood, parents and their parents, but also various cultures and life as a whole on the Planet Earth – it is all connected. Some will let their past influence them in a positive way, while others in a negative way. Knowing we came here to be on a chosen path, inside the Circles of Time, within The God Clock, choosing to learn and grow from our past might change the current moment, and while we are still on our chosen path, the circles in which we are experiencing that path, in this moment, may change, or take us on, or into something more positive. It could also simply be the same path, within the same circles, but a completely different view, or perspective of it. Again, knowing we connected to everything, including the Past and Future, it really shows us that NOW is much more important than we realize.

Why is there Time?

Why does time exist? The answer to that might be very simple – YOU are the reason and purpose time exists. Our path is defined by time and it is the journey we are on that began when our soul, or spirit manifested on the material plane setting in motion our own circles of time. On our path connecting with others circles of time and within time itself, inside the very essence, within the graceful, perfect mechanism – The God Clock. If you think about it, we and our journey through time, are literally woven into the very essence and fabric of time.

This Circle of Time

So now this book has completed its small, but important circle of Time. It has its own path to carry out, from its creation on through its own cycles and seasons and along the way making intersections with other circles, including yours and with others. Revealing and connecting us with the information left to us from the past within old calendars, structures and symbols that have been right in front of us this whole time.

Ancient time codes hidden within the heavens above, patiently waiting for the intersection to be made, leading us to the possible destination of all things, right from where we started – The God Clock.

The precise mechanism in which all time is unfolding – in a linear/sequential, cyclic, proportional and circular manner, interwoven and all at once. Built upon a simple concept of the relationship of a circle's relationship of its diameter to its circumference, quadrated, or thrown to the 4 directions. The gears moving and meshing perfectly together, made from the essence of time itself – the fabric of Time.

And through this intersection at this moment, might we realize our own journey through time within The God Clock and be more respectful, or aware of the choices we make and where those circles/connections might lead us. Recognizing the wisdom of the Native peoples that everything is a circle and that we are all connected – within the circle of truth.

And Finally – Will Time End?
The Big Sync

As illustrated in Chapter 4, within the Ancient Book of Time System, there are 90 plates, or mathematical quadrations of the perfect solar book (deck of cards), after 90 quadrations, the pages of the book fall back into perfect order, we then progress through these plates at various, seemingly separate rates closely tied to the natural cycles and seasons we experience here on Earth (all in Telios Proportions within the Star/Time Code.) But an interesting thing happens at age 90 – just as the 90 quadrations of the Perfect Solar Book fall back in to perfect order after 90 times and as mentioned, every progression from the Year on down, including the Seasonal, Lunar and Weekly, all sync or line up, just as they did when they started out the moment a person was born, exactly on Plate 90. After that, they simply start over.

So it has to be pondered, after millions, billions or even trillions of years, with all the circles of Time unfolding precisely within The God Clock, at some point, after all the possible circles and paths that can be taken and after all the connections and intersections can be made, will they too, all sync up? And if they do what will happen? Will time finally come to an end? Will the Universe collapse upon itself, only to then re-expand as some physicists believe? Will it just keep going and won't even be noticed? Or will the Universe and/or God simply take in a breath, ponder its magnificence, slowly exhale and start it all up again, one circle at a time.

The end of Time.......

About the Author

Thomas Morrell lives in the slow rolling hills of Eastern Iowa, near the Mississippi River. Living here in the Heartland has kept him both grounded and given him the ability to approach things in a very practical, no non sense manner.

Always being interested in how things work since an early age, his path in this life has led him on an unintended, incredible journey of discovery. While delving into ancient numeric, astronomical and ancient systems to look for possible answers, he saw a common denominator between them and started putting the pieces of the puzzle together.

This has led to important discoveries and understanding in regards to an ancient astronomical system, which he calls "The Ancient Book of Time." His discoveries include: the mathematical code on which the system unfolds, revealing 5 new charts or increments of time not known in modern times (there were originally only 3 without even a basis for 2 of them) – allowing the smallest calculations from 52 days, now down to the day! The Alphabet/Letter Key allowing the delineation of both names/words and numbers, as well as, many other important understandings within the System, not known before. He has written several books on the subject, with more to come.

All of this research and discovery has led to the writing and publishing of a remarkable new book on Time itself titled

The God Clock - a new look at the passage of time, where he proposes a new theory on how time unfolds in 4 different ways at the same time in the mechanism known as The God Clock.

When not pondering the great mysteries of life and our journey through it, Thomas spends his time building/paddling wood canoes and kayaks, designing and building his own archery equipment and still exploring the outdoors, just as he has done since a child. His unintentional quest to figure out how things work has never ended.

For more information, informative articles and books by the author visit theancientbookoftime.com

Printed in Dunstable, United Kingdom

91222395R00087